Applied Power Quality

Applied Power Quality

Analysis, Modelling, Design and Implementation of Power Quality Monitoring Systems

SARATH PERERA
Australian Power Quality and Reliability Centre, University of Wollongong, Wollongong, NSW, Australia

SEAN ELPHICK
Australian Power Quality and Reliability Centre, University of Wollongong, Wollongong, NSW, Australia

ELSEVIER

Elsevier
Radarweg 29, PO Box 211, 1000 AE Amsterdam, Netherlands
The Boulevard, Langford Lane, Kidlington, Oxford OX5 1GB, United Kingdom
50 Hampshire Street, 5th Floor, Cambridge, MA 02139, United States

ISBN: 978-0-323-85467-2

For information on all Elsevier publications
visit our website at https://www.elsevier.com/books-and-journals

Publisher: Charlotte Cockle
Acquisitions Editor: Rachel Pomery
Editorial Project Manager: Aleksandra Packowska
Production Project Manager: Prem Kumar Kaliamoorthi
Cover Designer: Vicky Pearson Esser

Typeset by STRAIVE, India

Contents

Acknowledgements

I wish to acknowledge my past and present colleagues at the Australian Power Quality and Reliability Centre (APQRC) who have inspired me to delve in to the quite fruitful subject of power quality. In particular, I wish to thank Emeritus Professor Vic Gosbell in that regard. I have been very fortunate to have brilliant postgraduate students who went along exploratory paths which helped me to open new ways of looking at power quality phenomena associated with voltage fluctuations and voltage unbalance. I wish to acknowledge the support of my colleagues from a number of CIGRE working groups who have acknowledged the work carried out at APQRC.

I have tremendous admiration of my long-standing colleague, Sean Elphick, who has been a backbone of APQRC. I also wish to acknowledge the support and insightful knowledge of my colleagues the late Associate Professor Phil Ciufo, Associate Professor Duane Robinson, Jason David and Gerard Drury of APQRC.

Sarath Perera

Firstly, I would like to thank my colleagues (past and present) at the Australian Power Quality and Reliability Centre, specifically and in no particular order: Professor Vic Gosbell, Associate Professor Duane Robinson, Associate Professor Phil Ciufo, Dr Vic Smith, Jason David, Gerrard Drury and Amin Rajabi. I have been fortunate in my career to have been able to work with such fine mentors and colleagues.

I'd also like to thank my co-author Professor Sarath Perera who has been a mentor to me for a long period of time.

Finally, I thank my family who have supported me through the process of compiling this book and perhaps have had to put up with a few too many weekends of dad in the study working on his book.

Sean Elphick

CHAPTER 1

Introduction to power quality in modern power systems

1.1 Introduction

In industrialised countries, electricity supply reliability, as measured by the frequency and duration of electricity supply interruptions, is now very high. Whilst reliability, or whether or not any level of voltage is present, may strictly be considered part of the umbrella of disturbances included in the broad definition of power quality, the fact that management of reliability is well understood and that reliability is of a high standard means that reliability and power quality are often considered to be separate topics. Given the aforementioned high levels of reliability, focus has moved to other areas of power quality where improvements can be made. Whilst many power quality concepts have existed for many years, in recent times there have been some very significant changes to the nature of the loads and generators that are connecting to electricity supply networks. Whilst these new loads and generators are not producing any new power quality disturbances, their characteristics and the methods by which they are being integrated into networks have resulted in power quality challenges and continued emphasis on the overall management of power quality from both a technical and regulatory perspective. Examples of recent power system changes that are directly relevant to power quality performance are as follows:

- There has been a proliferation of small-scale solar photovoltaic (PV) generation being integrated into low voltage (LV) networks. The majority of these systems are installed on household rooftops. As an example, Fig. 1.1 shows the trend of total installed capacity of solar PV in Australia over the past 20 years. The figure shows that take-up increased rapidly from 2010 onward and continues to increase rapidly. Integration of these distributed solar PV generators into low voltage electricity distribution networks has resulted in power quality challenges related to voltage regulation, voltage and current unbalance and to a lesser extent harmonic distortion.

Applied Power Quality
https://doi.org/10.1016/B978-0-323-85467-2.00006-8

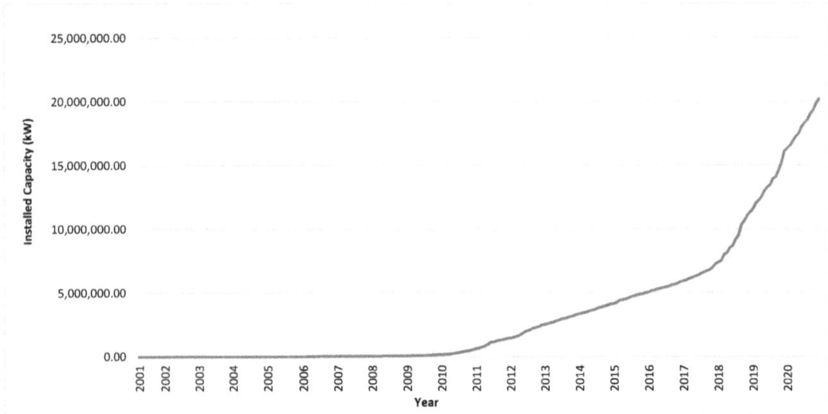

Fig. 1.1 Trend of installed solar PV capacity in Australia. *(Data from Australian PV Institute (APVI), Solar Map, Funded by the Australian Renewable Energy Agency, Webpage (Last Accessed 7 April 2021).)*

- Solar inverters and/or associated monitoring systems provide an indication of network voltage levels and in some cases, inverter systems are disconnecting from the network due to overvoltage protection settings. This has increased consumer awareness and visibility of network power quality and in turn resulted in a requirement for electricity network operators to take action to improve power quality.
- There has been a significant increase in the number of large renewable generators. These generators present power quality management challenges for the following reasons:
  - The location of the renewable energy resources that they capture is often in remote locations, and as such, there is limited transmission or subtransmission infrastructure. Consequently, many of these generators are being connected to weaker parts of networks making management of power quality more difficult.
  - The technologies used to integrate these generators with electricity transmission and distribution networks have performance and characteristics that are different to more traditional loads and generators. For example, inverters for large solar farms may inject harmonic orders which are atypical.
- There has been a continual shift in the characteristics of consumer appliances towards all devices being supplied by power electronics. The main reason for this shift is energy efficiency. Many devices which incorporate induction motors, for example, air conditioners, washing machines and

refrigerators, which once may have been direct-online start, are now more and more likely to include an inverter as the interface between the supply and the motor. In the case of air conditioners, it is now difficult to find a model that is not of the inverter type.

- The characteristics of lighting systems have completely changed over the past decade. In many countries, incandescent lamps, which are a linear load with few power quality implications, are being phased out. This is again for reasons of energy efficiency. In the first instance, incandescent lamps were replaced with compact fluorescent lamps (CFL); however, these have been replaced in turn by LED lighting systems. Whilst there is no argument that CFLs and LEDs use less energy than equivalent incandescent lamps, they are electronic devices which do emit harmonic distortion. These electronic lighting systems have also been observed to be more susceptible to power quality disturbances than incandescent lamps.

In addition to the changes to loads and generators specified above, it is highly likely that the number of electric vehicles and battery energy storage systems will increase over the coming years. Both of these technologies are interfaced with the wider electricity supply network through electronics and as such have the potential to both impact network power quality levels and/or be susceptible to power quality disturbances. Depending on their implementation, these technologies also have the potential to improve power quality.

1.2 What is power quality?

There are many definitions for power quality. The International Electrotechnical Commission (IEC) defines power quality as "characteristics of the electricity at a given point on an electrical system, evaluated against a set of reference technical parameters" [1] whilst the Institute for Electrical and Electronics Engineers (IEEE) defines power quality as "the concept of powering and grounding electronic equipment in a manner that is suitable to the operation of that equipment and compatible with the premise wiring system and other connected equipment" [2].

In simple terms, any deviations from a sinusoidal waveform of nominal voltage and frequency may be considered as a manifestation of a power quality disturbance. Power quality covers a wide range of phenomena that may be observed in power systems. The impacts of power quality range from relatively benign to catastrophic. By way of terminology, power quality phenomena are often called disturbances; however, the IEC also terms power quality phenomena parameters. For the purposes of this book, the term disturbance has been adopted.

Power quality disturbances can manifest in either voltage or current. In many cases, voltage disturbances are a result of the interaction between current and network impedance. For example, harmonic voltage distortion is caused by the interaction between the distorted current drawn by nonlinear loads and network impedance whilst voltage sags are caused by the interaction between fault current and network impedance. As a general principle, particularly under the IEC power quality management philosophy, network operators are responsible for maintaining acceptable power quality levels in voltage waveforms whilst customers are responsible for managing their power quality emissions at a level whereby they do not have an unacceptable impact on voltage.

1.2.1 Main power quality disturbances

This section provides a brief overview of the definition and impact of the main power quality disturbances. A comprehensive definition of each power quality disturbance is provided in the individual chapters dedicated to each disturbance.

1.2.1.1 Steady-state voltage variation

Voltage variation may be considered to be the most basic of power quality disturbances. The steady-state voltage magnitude is generally accepted to be the root-mean-square (RMS) value of the voltage waveform. This voltage magnitude will vary over time due to a range of factors including loading and network operation (e.g. transformer tap changer positions, connection or disconnection of capacitor banks).

The magnitude of the supply voltage is a key parameter for equipment operation. If voltage magnitudes are too low, equipment may not operate as intended or at all. If voltage magnitudes are too high, equipment may experience loss of life, consume additional energy or may fail.

1.2.1.2 Steady-state frequency variation

Frequency is another basic network operation parameter. In large interconnected networks, frequency is generally bound within tight limits and controlled through complex control strategies for generators. With the increase in nonsynchronous generation as well as movement towards higher numbers of stand-alone power supplies, maintenance of very tight frequency limits is becoming more challenging.

The main impact of variation in frequency appears to be implications for devices which count zero crossings. For these devices, variation in frequency can cause clocks to run fast or slow depending on whether the frequency is above or below the nominal value.

1.2.1.3 Unbalance

Unbalance, also termed imbalance, is defined by the IEC as a "condition in a polyphase system in which the rms values of the line voltages (fundamental component), and/or the phase angles between consecutive line voltages, are not all equal" [1]. Fig. 1.2 shows a visualisation of unbalance.

The impact of unbalance is mainly on three-phase devices. Perhaps the best known impact is that of voltage unbalance inducing counter-rotating magnetic fields in induction motors leading to additional heating which sometimes requires derating of the motor. Voltage unbalance may also cause three-phase devices to emit atypical harmonic distortion. In three-phase supply networks, unbalance will cause current to flow in the neutral conductor. In some cases, before unbalance was well understood, this was an issue if the neutral conductor was not sized appropriately. However, this issue is now well understood and wiring rules ensure that the neutral conductor is now appropriately sized.

1.2.1.4 Harmonic distortion

Harmonics are sinusoidal waveform components with frequencies that are integer multiples of the fundamental frequency. When harmonics are combined with the fundamental frequency the outcome is a distorted waveform. Harmonics distort the voltage or current waveform identically for each cycle. Some types of equipment (e.g. power electronic converters) will intrinsically draw distorted current made up of harmonics. The interaction

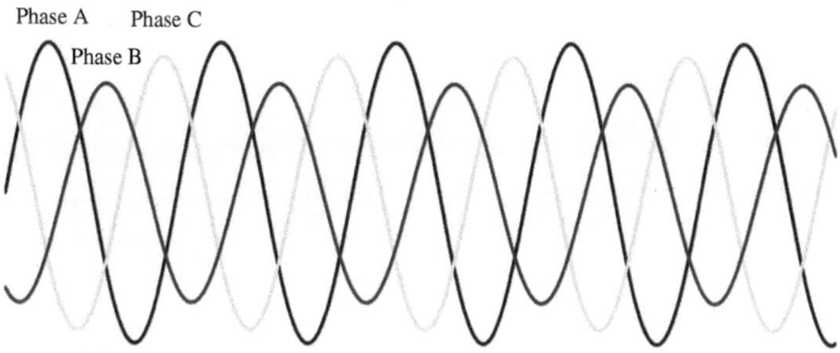

Fig. 1.2 Visualisation of unbalance.

Fig. 1.3 Example of waveform containing harmonic distortion.

between this distorted current and network impedance is the cause of distorted voltage waveforms and voltage harmonics. Fig. 1.3 shows an example of a waveform containing harmonic distortion.

Interharmonics are waveforms with frequency components that are not integer multiples of the fundamental frequency and originate from devices such as static frequency converters, cycloconverters and arc furnaces. The waveform distortion they produce is different for each half-cycle. Interharmonics are generally not of sufficient magnitude to be of concern. Fig. 1.4 shows an example of a waveform containing interharmonic distortion.

The impact of harmonic distortion can be quite wide and varied and includes

- Additional heating and losses in equipment and supply networks.
- Interference with control and communication systems.
- Catastrophic failure of equipment.

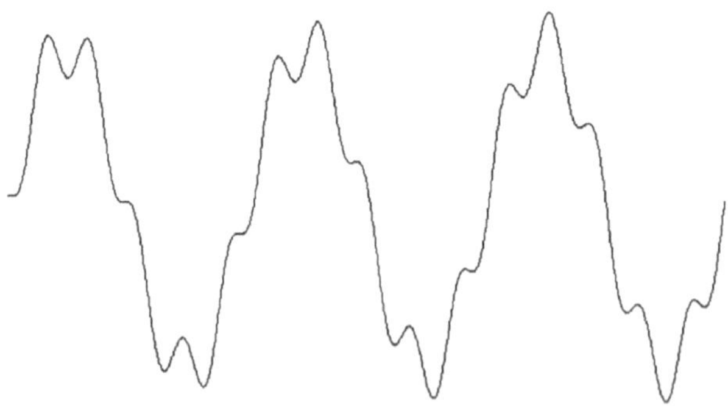

Fig. 1.4 Example of waveform with interharmonic distortion.

1.2.1.5 Voltage fluctuations and flicker

Voltage fluctuations and flicker are often grouped together; however, they are not equivalent. Voltage fluctuations are rapid changes in the magnitude of the voltage whilst flicker is an ongoing modulation of the amplitude of the voltage waveform envelope. In a simple sense, voltage fluctuations may be thought of as the source and flicker the outcome; however, not all voltage fluctuations lead to problematic flicker. The main impact of voltage fluctuations in isolation is dimming of lights and potential tripping of equipment. The impact of voltage fluctuations on equipment lifespan remains an area requiring further research. Isolated voltage fluctuations will not cause flicker; however, ongoing periodic voltage fluctuations are the source of flicker.

Ongoing voltage fluctuations can cause rhythmic changes in the output (flickering) of some lighting systems and this is the phenomena known as flicker. The IEC defines flicker as "impression of unsteadiness of visual sensation induced by a light stimulus whose luminance or spectral distribution fluctuates with time" [3]. The main impact of flicker is on humans where the ongoing changes in light output can cause a range of deleterious health impacts, ranging from inability to concentrate through to seizures in extreme cases. Fig. 1.5 shows a visualisation of the periodic amplitude modulation of the voltage waveform envelope that is the root cause of flicker.

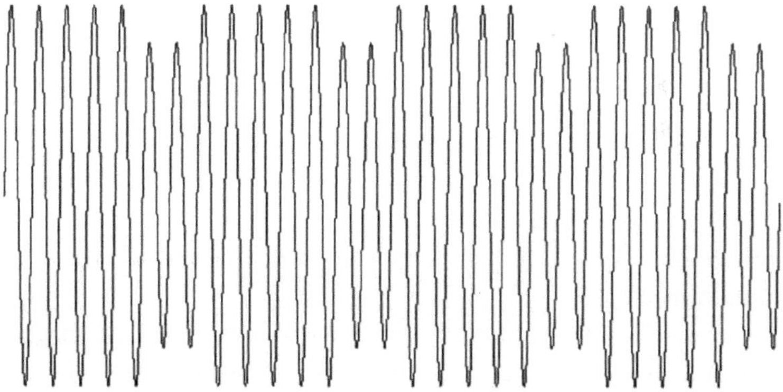

Fig. 1.5 Visualisation of the periodic amplitude modulation of the voltage waveform envelope that causes flicker.

1.2.1.6 Voltage sags

A voltage sag, also known as a voltage dip, is a short-term reduction of the voltage magnitude. Voltage sags are identified by threshold values of voltage magnitude. There is no consistent international definition of the voltage magnitude or time period that defines a voltage sag. The IEEE defines a sag as "a decrease in rms voltage to between 0.1 and 0.9 pu for durations from 0.5 cycles to 1 min" [4]. The IEC is much less prescriptive with respect to the magnitude of voltage and duration that apply to voltage sags, specifying a voltage dip (sag) as a "temporary reduction of the voltage magnitude at a point in the electrical system below a threshold" [5]. Fig. 1.6 shows an example of a voltage sag.

Voltage sags may well be the power quality disturbance with the greatest economic impact. Voltage sags can cause equipment to trip (cease operating). This tripping causes lost productivity and can lead to significant loss of production and raw materials in industrial plants.

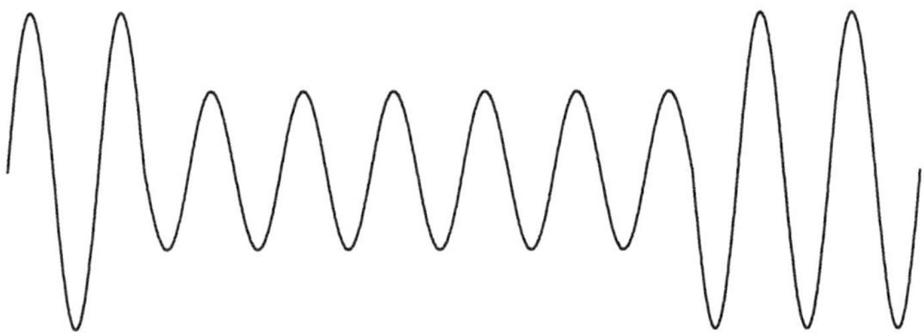

Fig. 1.6 Voltage sag.

1.2.1.7 Voltage swells

A voltage swell is a short-term increase in the voltage magnitude. In simple terms, voltage swells may be thought of as the opposite of voltage sags. However, measurement, analysis and reporting of voltage swells are not nearly as well advanced as that seen for voltage sags. This may be due to the fact that voltage swells are much rarer than voltage sags and do not have the same economic impacts.

Voltage swells are identified by threshold values of voltage magnitude. Similar to the case for voltage sags, there is no consistent international definition of the voltage magnitude or time period that defines a voltage swell. The IEEE defines a swell as "a swell is an increase in rms voltage above 1.1 pu for durations from 0.5 cycle to 1 min" [4]. Once again, the IEC is less prescriptive with respect to the magnitude of voltage and duration that apply to voltage swells, specifying a voltage swell as a "temporary increase of the voltage magnitude at a point in the electrical system above a threshold" [5]. Fig. 1.7 shows an example of a voltage swell.

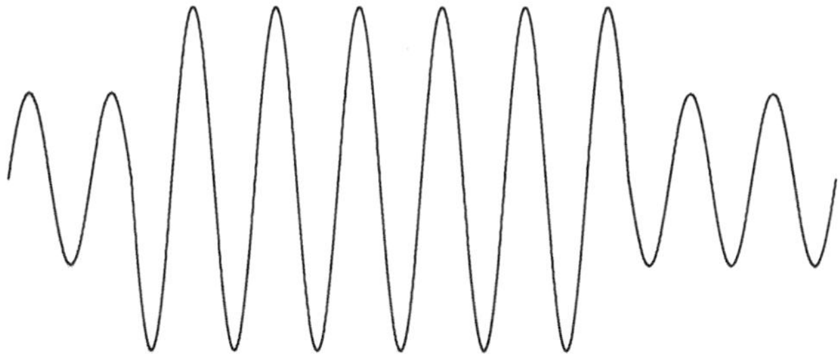

Fig. 1.7 Voltage swell.

1.2.1.8 Transients

A transient is a very short term, often subcycle, disturbance in the AC waveform that is evidenced by a sharp, brief discontinuity of the waveform. Transients are also known as spikes or surges and normally are on the line for only 1/1000th of a second or less (less than 1 millisecond). They can be from a few to 10,000 volts-peak above or below the voltage sine wave or data line signal. The impact of transients can range from tripping of equipment and disruption of control systems through to catastrophic failure.

There are two types of transients, namely, oscillatory or impulsive. Oscillatory transients are generally caused by energisation of loads with capacitor bank energisation being a well-known cause. Oscillatory transients are characterised by bidirectional excursions in the waveform followed by oscillation or ringing. Fig. 1.8 shows an example of an oscillatory transient.

Impulsive transients are generally caused by lightning. These transients are characterised by very short term and often very large unidirectional excursions of the waveform. Fig. 1.9 shows an example of an impulsive transient.

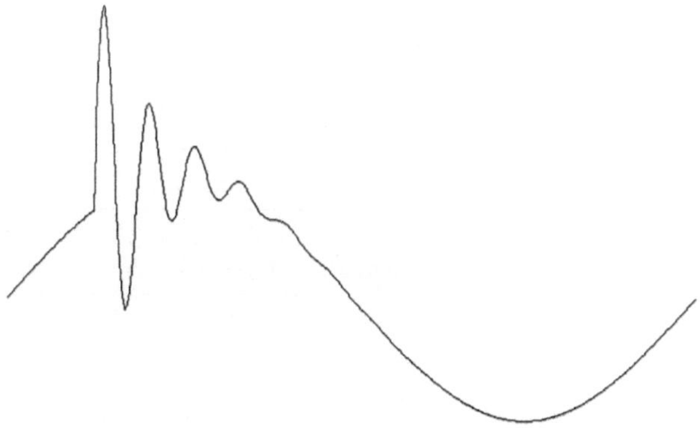

Fig. 1.8 Oscillatory transient.

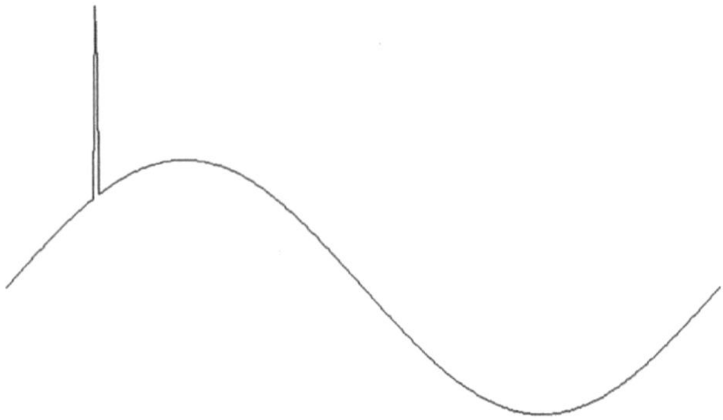

Fig. 1.9 Impulsive transient.

1.2.1.9 Characterising power quality disturbances

When considering monitoring, analysis and reporting of power quality disturbance levels, it is common to characterise power quality disturbances as either continuous or discrete phenomena.

Note: The IEC does not use the term continuous disturbance instead adopting the term quasisteady-state parameter; however, the term continuous disturbance is used in this book.

Continuous disturbances are those which are always present to some degree. Discrete disturbances occur as isolated disturbances over a few cycles potentially with a long interval before they are repeated. The concept of 'steady state' is not applicable to these disturbances since, if the term was to mean anything, it would be that they are absent. Table 1.1 presents the continuous or discrete classification for the most common power quality disturbances.

The philosophy for management, monitoring and analysis of continuous and discrete disturbances is very different, and as such, each disturbance classification needs to be addressed individually.

Table 1.1 Continuous and discrete disturbance classification.

Continuous disturbances (present every cycle)	Discrete disturbances (present in a small fraction of cycles)
Voltage variation	Sag/interruption
Unbalance	Swell
Harmonic distortion	Transient
Voltage fluctuations/flicker	

1.3 Power quality management philosophy

The overall aims of managing power quality disturbance levels within electricity supply networks are as follows:

- To ensure a safe and secure electricity supply.
- To ensure that all equipment connected to the electricity network operates as expected.
- To fully utilise the absorption capacity of the electricity network ensuring the optimum economic outcomes for consumers and network operators.

Management of power quality levels in electricity networks to ensure that all of the earlier mentioned objectives are met is achieved through a combination of equipment design guidelines along with measures undertaken to limit the magnitude of disturbances present in electricity. In most cases, there are standards and/or regulations that specify the methods that network operators should use to implement a power quality management philosophy.

In terms of equipment design there are standards which define a compatibility level. The IEC defines compatibility level as "the specified electromagnetic disturbance level used as a reference level for co-ordination in the setting of emission and immunity limits." The compatibility level is effectively the highest level of any disturbance that should be present in the electricity supply network. Equipment manufacturers use the compatibility level to design equipment such that the equipment immunity level is greater than the compatibility level, thus ensuring that equipment should operate as expected for any power quality disturbance level likely to be present in the supply voltage.

Once compatibility levels have been established electricity network planning methodologies must be implemented to ensure that power quality disturbance magnitudes present in electricity networks do not exceed the compatibility level. There are two aspects to this planning methodology:

1. Specification of planning levels—The IEC defines planning level as "level of a particular disturbance in a particular environment, adopted as a reference value for the limits to be set for the emissions from the installations in a particular system, in order to co-ordinate those limits with all the limits adopted for equipment and installations intended to be connected to the power supply system." The planning level effectively identifies the maximum disturbance magnitude that should be present at a particular location.

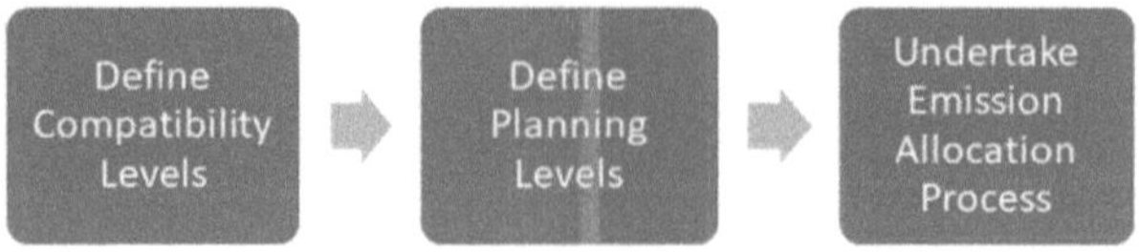

Fig. 1.10 Flowchart for implementation of a power quality management philosophy.

2. Allocation of emission levels to large loads—Emission level is defined by the IEC as "level of a given electromagnetic disturbance emitted from a particular device, equipment, system or disturbing installation as a whole, assessed and measured in a specified manner." For large installations connecting to medium voltage networks and above, it is typical to undertake an emission allocation process in order to limit the overall contribution that any particular installation (load or generator) may make to overall power quality disturbance levels through their power quality emissions. For low voltage networks, the power quality emissions of equipment and installations are generally controlled through standards which limit the magnitude of emission from individual pieces of equipment and it is less common for emission allocations to be provided for low voltage installations.

Based on the philosophies described before, Fig. 1.10 provides visualisation of the process for implementing a power quality management philosophy. Fig. 1.11 provides visualisation of the overall power quality management philosophy, that is, how compatibility levels, planning levels and equipment immunity work together in order to ensure appropriate power quality outcomes for all stakeholders. Network disturbance levels are due to the

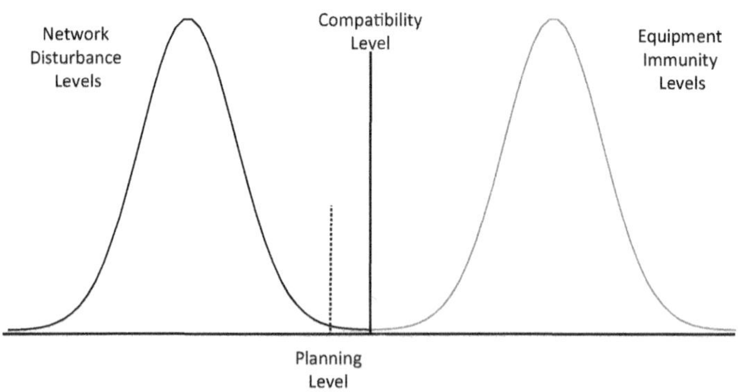

Fig. 1.11 Visualisation of power quality management philosophy.

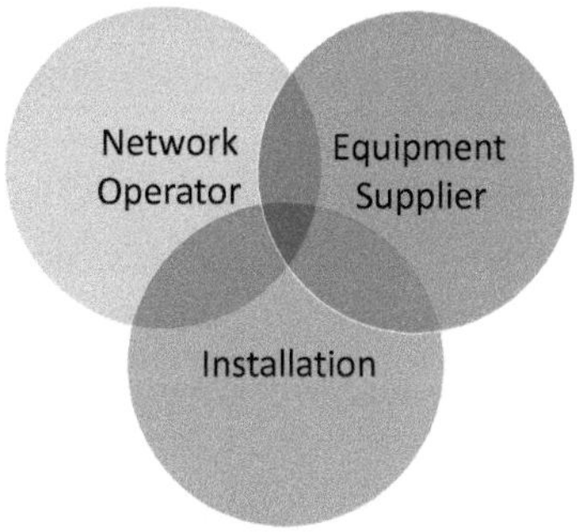

Fig. 1.12 Visualisation of overlapping responsibilities for network operators, equipment suppliers and installations for power quality management.

cumulative impact of all equipment connected to the network whilst equipment immunity levels are determined through type testing.

In order for the entire power quality management philosophy to be effective, a holistic approach must be adopted which links compatibility levels with planning levels and the emission allocation strategy. Each of these aspects must be well understood by the key stakeholders, namely, the network operator, equipment suppliers and installations. Fig. 1.12 provides visualisation of the overlapping responsibilities required for effective management of power quality.

1.3.1 Power quality standards

Whilst bespoke power quality management philosophy and standards exist for individual countries and even individual organisations, the predominant power quality management methodology and standards used or adopted in most locations are either those produced by the IEC or the IEEE.

1.3.1.1 IEC standards

The IEC has a highly structured suite of power quality standards for power quality which are contained within the umbrella of Electromagnetic Compatibility (EMC) and specifically the 61000 series. The IEC 61000 series of standards is published in separate parts according to the following structure:

1. General: fundamental principles, definitions
2. Environment: description and classification of the electromagnetic environment, compatibility levels
3. Limits: emission and immunity limits
4. Testing and measurement techniques: e.g. for assessing equipment immunity and for assessing disturbance levels

5. Installation and mitigation guidelines
6. Generic standards: general procedures for testing equipment
7. Miscellaneous

The nomenclature for the 61000 series of IEC standards is 61000.Part Number.Standard Number. For example, instrumentation standards are therefore numbered in the form IEC 61000.4.x whilst limits for network operators, installations and equipment are numbered IEC 61000.3.x.

Whilst the catch-all term standards are often applied to IEC documents, the IEC actually has three tiers of document as follows:

- International Standard (IS)—An international standard is a document that has been developed through the consensus of experts from many countries and is approved and published by a globally recognised body. It comprises rules, guidelines, processes or characteristics that allow users to achieve the same outcome time and time again.
- Technical Specification (TS)—A technical specification approaches an international standard in terms of detail and completeness but has not yet passed through all approval stages, either because consensus has not been reached or because standardisation is seen to be premature.
- Technical Reports (TR)—Technical reports focus on a particular subject and contain, for example, data, measurement techniques, test approaches, case studies, methodologies and other types of information that are useful for standards developers and other audiences. They are never normative. Technical reports typically have not achieved the consensus required for publication as an international standard.

1.3.1.2 IEEE standards

The IEEE has a somewhat less structured approach to standardisation than the IEC. IEEE standards also tend to include much more 'tutorial' or informative type information than IEC standards. The IEEE standard for management of harmonics is IEEE Std 519 "IEEE Recommended Practices and Requirements for Harmonic Control in Electrical Power Systems." IEEE Std 1159 is "IEEE Recommended Practice for Monitoring Electric Power Quality" provides recommendations for monitoring of power quality.

IEEE colour book series also provide useful guidance for management of power quality, especially the following:

- IEEE Std 142-2007 "IEEE Recommended Practice for Grounding of Industrial and Commercial Power Systems," also known as the Green Book.

- IEEE Std 1100-2005 "IEEE Recommended Practice for Powering and Grounding Sensitive Electronic Equipment," also known as the Emerald Book.
- IEEE Std 446-1995 "IEEE Recommended Practice for Emergency and Standby Systems for Industrial and Commercial Applications," also known as the Orange Book.

1.4 Overview of contents

This book has been written to specifically assist practitioners working in power quality; however, a range of individuals working in the electrical supply industry may find aspects of the book useful. The book is designed to assist readers to have a better understanding of power quality and to explain and simplify complex aspects of their jobs. The content presented has been collated to provide a one-stop guide to broad management of power quality from equipment behaviour through to network operation and planning. Evaluation of the performance and trends in modern equipment are provided. Where relevant, comprehensive guidance on the standards applicable to management of power quality is also provided. The theoretical material presented in the book is supported by the latest research as well as laboratory and field measurements of equipment and network data. The book may also be useful in assisting with the design of power quality allocation, monitoring and assessment systems. This book is organised as follows:

1.4.1 Chapter 2: Steady-state voltage in low voltage networks

This chapter discusses the relevance of managing steady-state voltage magnitudes in low voltage networks. The chapter details the impact of the proliferation of small-scale renewable energy devices on voltage and the strategies that can be used to manage voltage regulation. The chapter also examines the impact of supply voltage on equipment performance, including findings related to loss of equipment life.

1.4.2 Chapter 3: Impact and management of power system voltage unbalance

This chapter discusses voltage unbalance including its causes and effects including impact on equipment behaviour. The chapter contains outcomes of the latest research into the impact of unbalance on induction motor performance. The chapter also provides some details of the economic impact of voltage unbalance on consumers.

1.4.3 Chapter 4: Impact and management of power system harmonics

This chapter describes power system harmonics including their causes (with the focus being on modern equipment) and effects including impact on equipment behaviour. The chapter also provides an in-depth explanation of the complex IEC guidelines which are used for management of harmonics, including application of the standards to large renewable energy generators (currently a topic of much conjecture). The limitations associated with the existing approaches used to study harmonic compliance in networks are also addressed.

This chapter also examines the topical subject of high-frequency harmonic emissions from small-scale renewable generators.

1.4.4 Chapter 5: Impact and management of power system voltage fluctuations, flicker and rapid voltage changes

This chapter details the causes and effects of voltage fluctuations, flicker and rapid voltage changes, including impact on equipment behaviour. The chapter also examines the IEC guidelines that are applicable to voltage fluctuations and flicker, including explanation and simplification of the emission allocation for flicker and rapid voltage changes.

1.4.5 Chapter 6: Impact and management of power system voltage sags

This chapter examines the causes and impacts of voltage sags, including the economic impact of voltage sags on consumers. Modern methods of limiting and managing voltage sag behaviour are also examined.

1.4.6 Chapter 7: Implications of equipment behaviour on power quality

This chapter investigates the behaviour of modern equipment as it pertains to power quality. In the past decade there have been significant changes to the operating characteristics of devices connected to electricity distribution systems. Major changes include the movement to high efficiency lighting loads (i.e. LED lamps), the move towards power electronic front ends on most household appliances (e.g. washing machines and refrigerators) and the proliferation of renewable energy technologies at both small and large scale. The chapter will highlight the major differences between performance of modern equipment compared to traditional concepts with respect to equipment performance.

1.4.7 Chapter 8: Power quality monitoring, data analysis and reporting

This chapter focuses on power quality monitoring, data analysis and reporting. Methodologies for both reactive (i.e. in response to complaints or faults) and proactive (i.e. continuous monitoring to establish baseline performance) will be discussed. Topics discussed in the chapter include

- Rationale for PQ monitoring.
- PQ monitoring instruments.
- Analysis and reporting methodologies including indices.
- Standards for power quality monitoring.
- Transducers for power quality monitoring.

References

[1] IEC, IEC 60050, International Electrotechnical Vocabulary (IEV), 1990.

[2] IEEE, IEEE 100™, The Authoritative Dictionary of IEEE Standards Terms, seventh ed., 2000.

[3] IEC, Electromagnetic compatibility (EMC)—Part 3-7: Limits—Assessment of emission limits for the connection of fluctuating installations to MV, HV and EHV power systems, IEC TR 61000-3-7, 2008.

[4] IEEE Recommended Practice for Monitoring Electric Power Quality, IEEE Std 1159-2019 (Revision of IEEE Std 1159-2009), 2019, pp. 1–98.

[5] IEC, Electromagnetic Compatibility (EMC)—Part 4-30: Testing and Measurement Techniques—Power Quality Measurement Methods, 2008.

CHAPTER 2

Steady-state voltage in low voltage networks

2.1 Introduction

Steady-state voltage is probably the most fundamental of power quality disturbances and perhaps the most overlooked in terms of assessment and impact. For the purposes of this chapter, steady-state voltage is defined as voltage magnitudes equal to or relatively near to the system nominal voltage (i.e. this chapter is not concerned with voltage sags or interruptions, which are dealt with in Chapter 6 or transients which are dealt with to some degree in Chapter 7) and the magnitude considered is the root mean square (RMS) value of the voltage waveform. The emphasis in this chapter is also on the management and impacts of steady-state voltage magnitudes in low voltage (LV) networks as this is where the majority of customers and their equipment are connected. There is obviously a very strong relationship between upstream voltages and LV voltage magnitudes and this is dealt with where appropriate.

Steady-state voltage is unique amongst power quality disturbances insomuch as the optimum value is not zero. Instead, supply voltage magnitude must be maintained within a band suitable for equipment to operate efficiently and as expected. Generally, equipment operates at highest efficiency closest to the rated voltage (which is often the network nominal voltage). Equipment may not operate as intended outside of the nominal voltage range. Equipment loss of life and/or failure is likely if voltage magnitudes are allowed to become too high and equipment may not operate as intended if voltage magnitudes are allowed to become too low.

As noted previously, this chapter primarily examines supply voltage magnitude in LV networks. Standards for voltage and nominal voltage used throughout the world are described first. The principles of voltage regulation are then introduced, including voltage regulation with and without distributed generation. The impact of supply voltage magnitude on different equipment types as well as consumer electricity usage is then discussed. Finally, methods which can be used to improve unacceptable levels of voltage regulation in LV networks are described.

Applied Power Quality
https://doi.org/10.1016/B978-0-323-85467-2.00003-2

2.2 Voltage standards

2.2.1 Nominal voltages

There are two main nominal voltage magnitudes used across the world. These are 220/230V which is used in Europe, Australasia, Africa and the southern parts of South America and 110/120V which is the nominal voltage in North America and the northern parts of South America. There are also two nominal frequencies used worldwide—60Hz for those countries with a 110/120V nominal voltage and 50Hz for those countries with a 220/230V nominal voltage. Fig. 2.1 shows worldwide nominal voltages.

The network design philosophies for the two nominal voltages are quite different. LV networks with a nominal voltage of 110/120V generally have shorter line lengths (feeders) than those with nominal voltages of 220/230V and hence many more distribution transformers, with smaller ratings, are required.

2.2.2 Standards

Standards for voltage magnitude identify the nominal voltage and allowable range for which network operators must design and operate their networks. These standards also prescribe the tolerance boundaries to which equipment manufacturers should design and construct equipment. Table 2.1 presents a summary of standards for voltage magnitude across the world.

In practice, network operators generally have some scope to set their voltage range and in some cases an asymmetrical range may be preferred. As an example of nominal voltage and allowable range, Fig. 2.2 shows an example

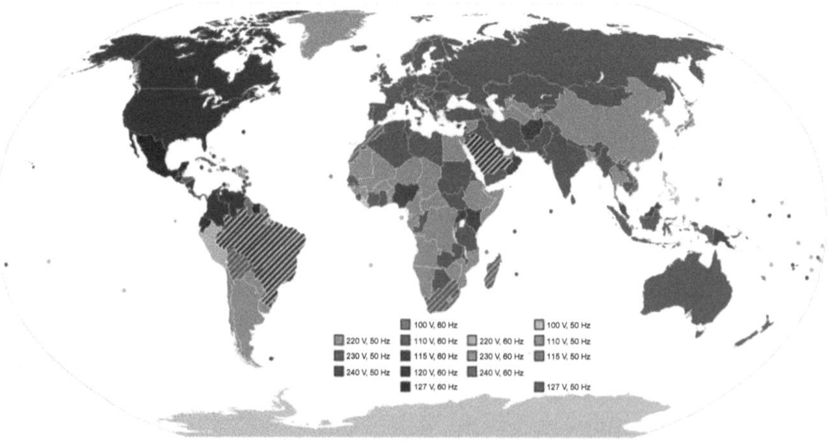

Fig. 2.1 Worldwide nominal voltages.

Table 2.1 Low voltage limits for different international standards.

Standard	Limits	Remarks
IEC 60038 Standard voltage	Nominal voltage ±10%	• LV nominal voltage is o 230 V line-to-neutral (L–N) o 400 V line-to-line (L–L) • Does not provide assessment criteria for voltage measurements
ANSI Standard. C84.1 Electric power systems and equipment—voltage ratings (60 Hz)	Range A: nominal voltage ±5% Range B: 91.7% for minimum voltage, 105.8% for maximum voltage	• Range A is optimal voltage range • Range B is acceptable but not optimal • The nominal LV voltage range is between 120 V and 600 V
EN 50160 Voltage characteristics of electricity supplied by public distribution systems	Nominal voltage ±10%	• Nominal voltage is o 230 V L–N o 400 V L–L • Assessment is based on the 95th percentile value over a 1-week period
AS 61000.3.100	Nominal voltage + 10%/−6%	• Nominal voltage is o 230 V L–N o 400 V L–L • 'Power quality' standard for voltage compliance. Assessment is based on 99th and 1st percentile voltage measurements over a 1-week period

of the nominal voltage (dotted black line at 230 V) and allowable voltage range for a network with a symmetrical voltage range of 230 V ± 10% and a system with an asymmetrical voltage range of 230 V + 10%/−6%. For these two systems, voltages between 207 V and 253 and 216 V and 253 V, respectively, would be considered to be within the allowable range.

2.3 Equipment response to voltage magnitude

Voltage standards are necessary to allow manufacturers to design equipment that will operate as expected when exposed to voltage magnitudes within the acceptable range. In some cases variations in voltage magnitudes will change

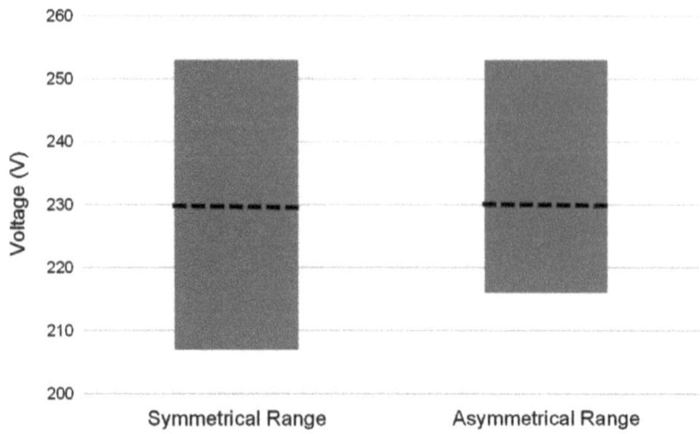

Fig. 2.2 Example of nominal voltage and voltage range (symmetrical and nonsymmetrical) for a 230V nominal system.

how a device behaves electrically. These are important factors to consider as one must balance the inherent cost of suitable supply voltage magnitude and the impact of non-nominal voltage levels on connected customers.

In simple terms loads may be characterised as either constant impedance (Z), constant current (I) or constant power (P). The relationship between supply voltage magnitude and current drawn for each of these load classifications is as follows:

- Z type loads—current will increase as voltage magnitude increases and vice versa.
- I type loads—current is not impacted by supply voltage magnitude.
- P type loads—current will increase as voltage magnitudes decreases and vice versa.

The previously mentioned relationships may have significant implications for electricity network operators. Where high levels of P type loads exist, reductions in voltage will lead to increased current. Such a scenario may have implications on the thermal rating of network assets such as switchgear, transformers and conductors. For high levels of Z type load, demand will increase as voltage increases and this again may have implications for network ratings. For a scenario in which a dominance of I type loads exist, the magnitude of current drawn is generally not impacted by a variation in voltage magnitudes.

2.3.1 Impact of voltage magnitude on equipment operation

The magnitude of supply voltage can have impacts on equipment operation, demand and lifespan. The general implication of operating at voltage magnitudes at or above the upper end of the allowable voltage range is as follows:

- Increased losses
- Reduced equipment lifespan
- Catastrophic failure via insulation breakdown or dielectric stress
- Increased energy consumption
- With respect to distributed generation, inverter protection systems will operate to disconnect the generation system once the voltage reaches the trip threshold. This reduces the amount of renewable energy generated and also leads to customer complaints to the network operators due to reduced revenue from feed-in tariffs.

Whilst for voltage magnitudes at or below the lower end of the voltage range implications include:

- Equipment fails to operate as intended and/or trips
- Inability to start motors and/or motor stalling
- Lower efficiencies
- Less hold-up time (i.e. ability to operate as expected or at all) when exposed to voltage sags
- Renewable energy generation will also trip if voltage levels become too low

Noting that the content in this section only considers voltage levels within the allowable range, i.e. not over- or undervoltage scenarios, further detailed discussion regarding the impacts of supply voltage magnitude on selected equipment types is given below.

2.3.1.1 Induction motors

The impact of supply voltage magnitude on induction motors is dependent on a number of factors, including the motor load and the motor load characteristics (e.g. pump or fan).

Voltage magnitudes below the nominal value will result in less starting torque or an inability to start in extreme cases. Heavily loaded induction motors approximate a constant power load and thus, reduced operating voltage will lead to increased current. The increased current associated with voltage magnitudes below nominal values leads to additional heating, commensurate temperature rise and reduced efficiency. Over the longer term, increased operating temperature will accelerate insulation breakdown and result in loss of life.

Lightly loaded induction motors can be approximated to a constant impedance load and there are more undesirable implications associated with overvoltage as opposed to undervoltage. For a lightly loaded induction motor, overvoltage will lead to increased current and in turn increased

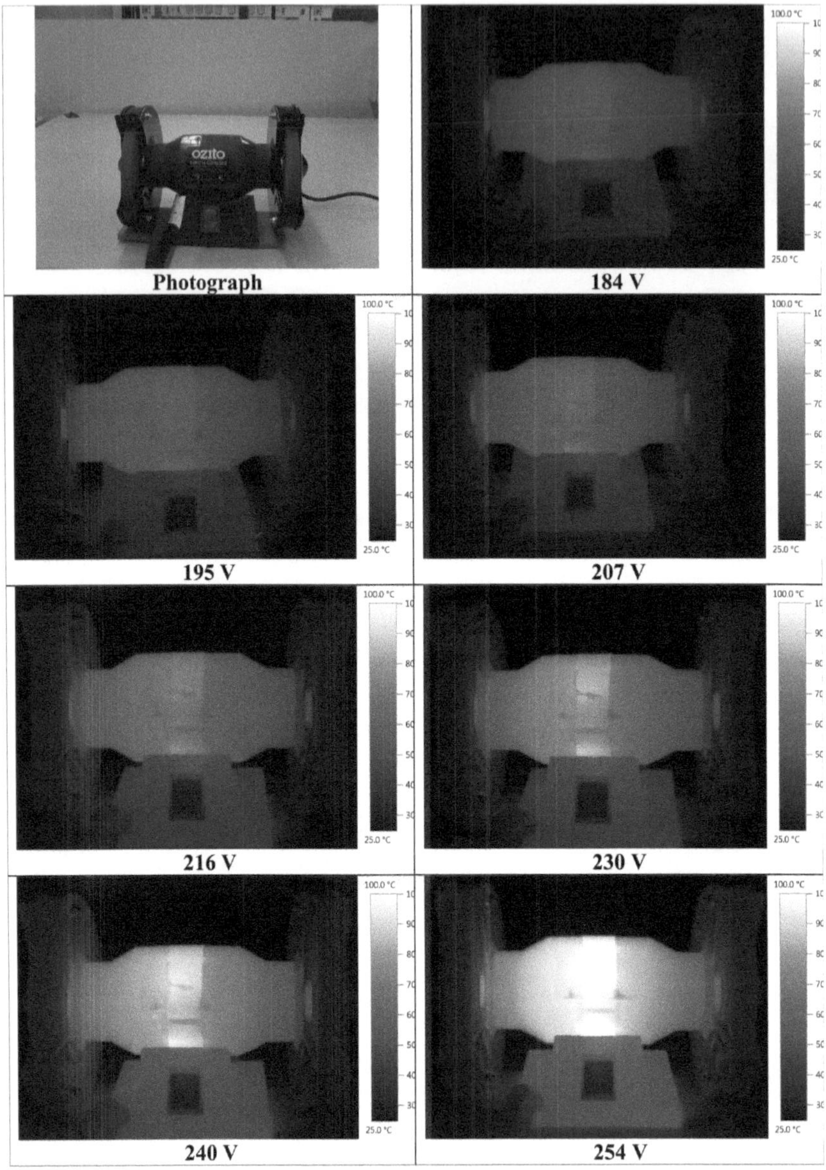

Fig. 2.3 Impact of supply voltage on operating temperature changes for a bench grinder.

heating as well as increased power consumption. As an example, Fig. 2.3 shows a thermal image of a bench grinder after two hours operating at each of the voltage magnitudes shown in the figure. The increase in operating temperature as supply voltage magnitude increases is clearly identifiable.

2.3.1.2 Lighting

Incandescent lamps

For incandescent lamps, the lamp filament can be approximated as a constant impedance load. The light output is directly proportional to the magnitude of the supply voltage, i.e. increasing supply voltage magnitude will increase light output. However, there is a very strong relationship between supply voltage and lamp life for incandescent lamps. For a sustained supply voltage magnitude V near the rated voltage of the lamp:

- *Light* output is approximately proportional to $V^{3.4}$
- *Power* consumption is approximately proportional to $V^{1.6}$
- *Lifetime* is approximately proportional to V^{-16}
- *Colour temperature* is approximately proportional to $V^{0.42}$

This means that a 5% reduction in operating voltage will more than double the life of the bulb at the expense of reducing its light output by approximately 16%. Conversely, a 5% increase in operating voltage above the rated voltage will halve the lifetime of the lamp, albeit with a higher light output as a trade-off. As listed before, variations in voltage magnitude will also cause the lamp to use more or less energy for higher or lower input voltage levels, respectively.

Tubular fluorescent lamps

The relationship between current and supply voltage magnitude for tubular fluorescent lamps with simple electronic or inductive ballast can also be approximated to the behaviour of a constant impedance load, i.e. current draw will increase as supply voltage increases. Fig. 2.4 shows the relationship between

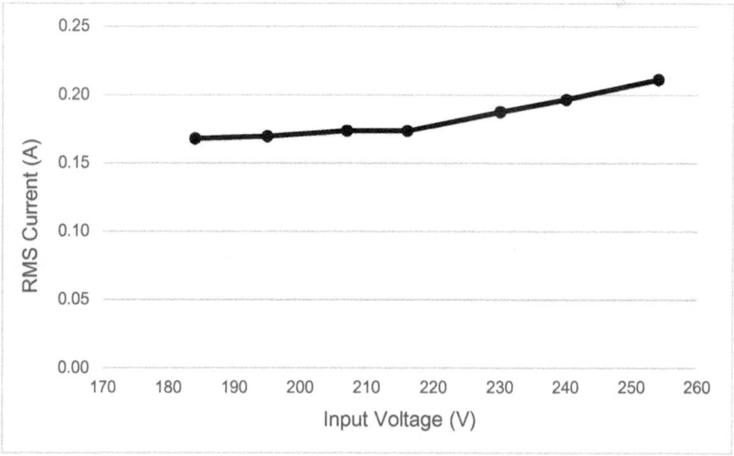

Fig. 2.4 Relationship between supply voltage magnitude and RMS current for a tubular fluorescent lamp.

rms current and input voltage magnitude for a tubular fluorescent lamp. Above 220 V, there is a very perceivable increase in current as voltage increases.

Modern lighting technologies

For more modern lighting technologies such as compact fluorescent lamps (CFLs) and light emitting diodes (LEDs), which are supplied by electronic circuits, the response of the device to varying input supply voltage magnitude may be more complex. Many CFLs can be approximated to a constant current load whilst LED lamps appear to operate more like a constant power load. Fig. 2.5 shows the relationship between input voltage magnitude and RMS current for a bank of CFL and LED lamps (12 of each). The constant current and constant power type characteristic of each lamp technology can clearly be observed.

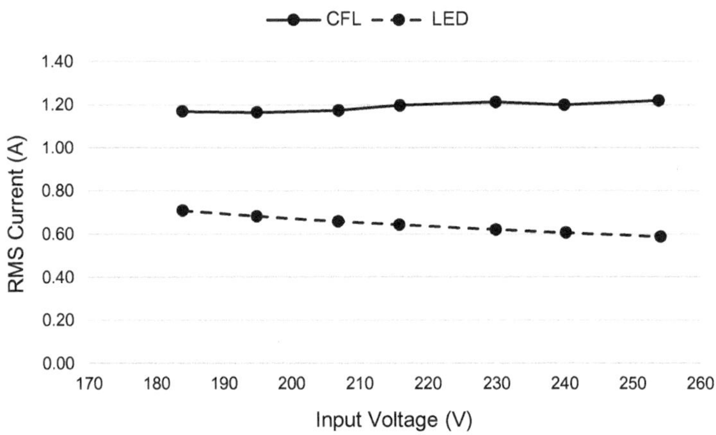

Fig. 2.5 Relationship between supply voltage magnitude and RMS current for a bank of compact fluorescent and LED lamps.

Capacitor/reactor banks

Capacitor and reactor banks are effectively a constant impedance (Z) type load. The rated reactive power will be achieved when the bank is supplied at rated voltage. When supplied at voltage below the rated voltage, the bank will supply less than rated reactive power. Conversely when supplied at voltages above the rated magnitude, the bank will supply more than the rated reactive power.

In the case of capacitors, these devices can be highly sensitive to overvoltage situations due to the insulation stresses. There is good evidence to show that capacitor lifespan can be significantly reduced if it is supplied with voltage magnitudes above the rated value. As an example, Fig. 2.6 shows the

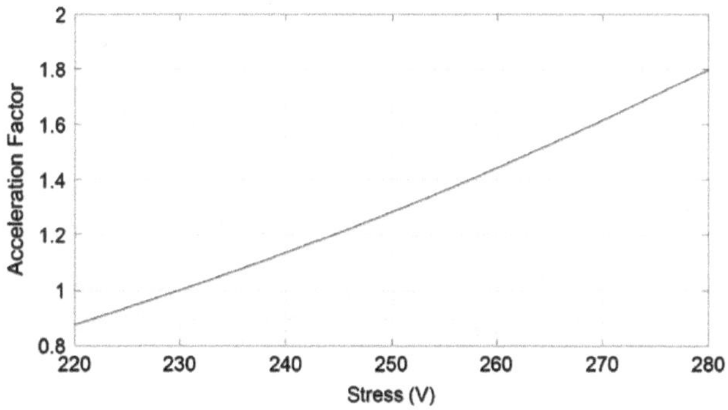

Fig. 2.6 Capacitor loss of life acceleration factor due to input voltage [1].

loss of life acceleration factor for a capacitor. The curve was derived using accelerated ageing testing in a laboratory. The graph indicates that if supplied at 250 V, the capacitor will age 1.25 times faster than if supplied at 230 V. If supplied at 280 V, the capacitor ages 1.8 times faster than if supplied at 230 V.

Transformers

The impact of the applied voltage magnitude on transformers is complex; there are losses associated with the magnetic circuit along with voltage-dependent load losses. That is, for a fixed tap transformer, as the applied primary voltage is increased, the losses in the transformer will increase as a consequence of the higher voltage, but also potentially in reaction to the response of the load(s) connected to the secondary winding of the transformer. For example, if the load consisted primarily of resistive loads, an increased operating voltage would lead to increased load current, thereby increasing load losses.

If the load is removed from the transformer, then it is possible to model the influence of the applied voltage on the transformer no-load losses. This situation is reflected in Fig. 2.7.

The transformer may require derating as the level of supply voltage increases. Conversely, as the supply voltage decreases, the no-load losses also decrease. Caution must be used when applying this model due to the high dependence of load behaviour in regard to on-load losses.

2.3.2 Impact of voltage magnitude on consumer electricity bills

The magnitude of the supply voltage can have a significant impact on consumer electricity bills depending on the characteristics of the load. It is generally accepted that consumer bills will be higher when the consumer is supplied at a higher voltage magnitude.

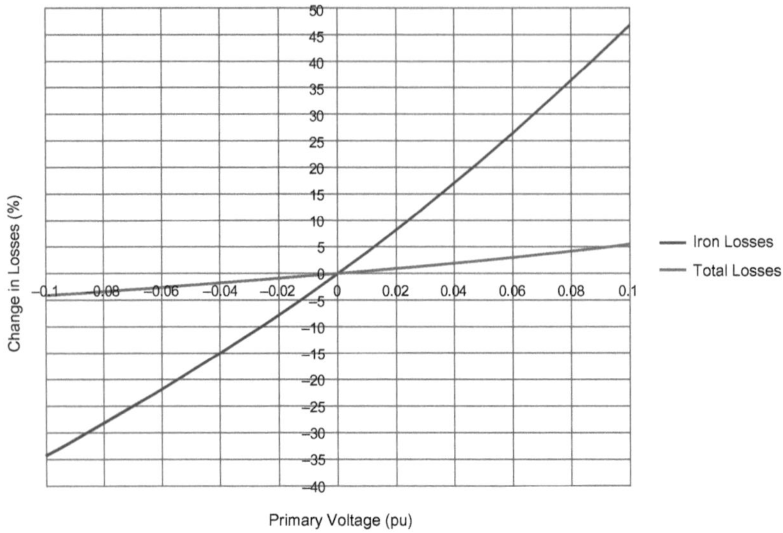

Fig. 2.7 Change in transformer losses versus applied voltage.

2.3.2.1 Conservation voltage reduction

Reducing voltage in order to reduce energy consumption forms one part of an energy reduction principle called conservation voltage reduction (CVR). The general definition of CVR factor is the ratio between the percentage change in value of the parameter of interest and the percentage change in voltage. For example, the CVR factor for active power (*P*) is calculated as follows:

$$\mathrm{CVR}_p = \frac{\Delta P(\%)}{\Delta V\ (\%)}$$

For this example a 1% change in voltage will produce a change in power equivalent to the calculated CVR factor.

There is significant evidence to suggest that a reduction in voltage level will lead to a reduction in domestic energy consumption. This concept is not new as evidenced by studies such as [2,3] from the 1980s. Recent studies such as [4,5] conducted in the USA have shown that reducing voltage levels will reduce energy consumption. For the study presented in [4] a widespread 5% reduction in voltage across an entire distribution network led to a 2% reduction in energy consumption, resulting in a CVR factor of 0.4. The study detailed in [5] found the following:

- "CVR provides peak load reduction and annual energy reduction of approximately 0.5%–4% depending on the specific feeder."

- "When extrapolated to a national level, it can be seen that a complete deployment of CVR, 100% of distribution feeders, provides a 3.04% reduction in annual energy consumption."
- "If deployed only on high value distribution feeders, 40% of distribution feeders, the annual energy consumption is still reduced by 2.4%."

The study presented in [6] was conducted using Australian appliances, voltage levels and distribution network characteristics. The study concluded that an active power reduction of 1.05% and reactive power reduction of 2.28% could be achieved for every 1% reduction in source voltage magnitude. The simulation results illustrated that CVR implementations could be effective in Australian networks to reduce demand and energy.

Table 2.2 presents CVR factors for common domestic appliances for active power (CVR_P) and energy (CVR_E) which have been calculated across a voltage range of 184 to 254V. Positive CVR factors indicate that the appliance uses less active power at 184V compared to 254V, whilst the reverse is true for negative CVR factors. The following provides a general summary of CVR factor performance as presented in Table 2.2.

- CVR factors for both power and energy for electronic appliances within the audio-visual group, PC and LED lamps were small indicating constant active power and energy operation.
- Power and energy CVR factors for the inverter air conditioner were also relatively small.
- Reasonably high power and energy CVR factors were observed for appliances with rotating motors, including vacuum cleaners, bench grinders, washing machines, fans and to a lesser extent pool pumps.
- Reasonably high power and energy CVR factors were observed for the majority of heaters; however, reduced heat output can be expected as supply voltage is reduced.
- Reasonably high power and energy CVR factors were observed for the CFL and fluorescent lights; however, reduced light output may be expected as supply voltage is reduced, especially for the fluorescent lamp.
- Reasonably high power and energy CVR factors were observed for microwave ovens.
- The portable air conditioners had modest CVR_P and CVR_E factors.
- Three of the four refrigerators had modest CVR_P and CVR_E factors. The fourth had a reasonably high CVR_P factor but a very small CVR_E factor.
- High CVR_P factors were observed for appliances used to heat water; however, the CVR_E factors for these appliances are low.

Table 2.2 Active power and energy CVR factors for a range of common domestic appliances.

Equipment type/specific devices	CVR_P	CVR_E
Kettles		
• Kettle 1		
• Kettle 2	2.37	−0.08
	2.31	−0.06
Audio visual		
• Television 1		
• DVD player 1	0.00	0.00
• DVD player 2	0.04	0.04
• Television 2	0.03	0.03
• Gaming console	0.01	0.01
• Clock radio	0.00	0.00
	2.44	2.44
Information technology (IT)		
• Desktop PC		
• Laptop PC	0.11	0.11
• PC monitor	2.44	2.44
	0.00	0.00
Heaters		
• Fan heater		
• Ceramic fan heater	2.28	2.28
• Heater 1	1.49	1.49
• Heater 2	0.33	0.33
• Bar element heater	2.20	2.20
• Oil column heater 1	2.37	2.37
• Oil column heater 2	1.48	1.48
	0.16	0.16
Lighting		
• LED lamps		
• CFL lamps	−0.04	−0.04
• Fluorescent tubes	0.99	0.99
	3.82	3.82
Microwaves		
• Microwave 1 (800 W)		
• Microwave 2 (1100 W)	2.78	4.10
	0.90	0.93
Fans		
• Pedestal fan 1		
• Pedestal fan 2	1.88	1.88
• Desktop fan	0.10	0.10
• Exhaust fan	2.06	2.06
	2.55	2.55

Table 2.2 Active power and energy CVR factors for a range of common domestic appliances—cont'd

Equipment type/specific devices	CVR_P	CVR_E
Bench grinders		
• Bench grinder 1	3.25	3.25
• Bench grinder 2	3.18	3.18
Vacuum cleaners		
• Vacuum cleaner 1	2.27	2.27
• Vacuum cleaner 2	2.26	2.26
Air conditioners		
• Portable air conditioner 1	0.50	0.49
• Split system inverter air conditioner	−0.03	−0.06
• Portable air conditioner 2	0.20	0.20
Refrigerators		
• Refrigerator 1	0.37	0.45
• Refrigerator 2	0.16	0.38
• Refrigerator 3	0.22	−0.01
• Refrigerator 4	1.24	0.03
Pool pumps		
• Pool pump 1	0.46	0.20
• Pool Pump 2	0.37	0.42
Washing machines		
• Washing machine 1 (new)	0.87	1.15
• Washing machine 2 (old)	1.42	1.60
Electric storage hot water heaters		
• Hot water heater 1	2.35	−0.11
• Hot water heater 2	2.30	0.08

Whilst CVR studies show the potential of reducing voltage across large distribution networks, they do not ensure that an energy saving will be achieved at a single domestic premises or, if an energy saving can be achieved, they give no indication of the magnitude of this saving. The effectiveness of any CVR implementation will principally be dependent on:

- The existing voltage level at the site due to the fact that the voltage can only be reduced so far before equipment will fail to operate correctly.
- The appliance mix at the site. Some appliances will use less energy for a reduced input voltage whilst others may use the same and a small number may use more. In order to understand how voltage reduction can be effective in reducing domestic energy consumption, it is necessary to understand the characteristics of all domestic appliances.

2.4 Causes of steady-state voltage variation

There are a range of factors which can cause variation in steady-state voltage magnitudes including to levels which are outside of the allowable range. The following list provides a summary of the main causes of variation in steady-state voltage magnitudes:

- **Overloading**—the interaction between the load current and network impedance at overloaded sites can cause unacceptable levels of voltage drop and result in voltage magnitudes below the allowable operating range.
- **Distributed energy resources**—as described in Section 2.5.2, distributed energy resources will always cause some degree of voltage rise. If either the rating of the distributed energy resource or the impedance of the network is too large, the amount of voltage rise may cause voltage magnitudes to exceed applicable limits.
- **Poor planning/coordination**—proper voltage regulation involves coordination between the upstream voltage (often determined by an on-load tap changer at the zone substation) and the off-load tap position of the distribution transformer. Improper tap positions on the distribution transformer can cause unacceptable voltage magnitudes.
- **Loss of voltage control at zone substation**—occasionally, voltage regulation systems at the zone substation may fail or operate incorrectly, this can lead to sustained unacceptable voltage magnitudes across large areas of the network.
- **Broken neutral conductor**—in unbalanced networks, a broken neutral conductor can cause significant variations in phase voltage magnitudes.
- **High impedance connections**—high impedance network connections which can be observed in bonds, joints and other connections can lead to unacceptable voltage drop.

2.5 Principles of voltage regulation in LV feeders

The various nominal voltages applied in different parts of the world were discussed before. It is important to note that whilst the nominal voltage may be the 'target' or most desirable voltage and the magnitude at which equipment is designed to operate most efficiently, it is neither technically nor economically feasible for the network operator to supply the entire LV network at the nominal voltage. This is due to the fact that there will always be variation in voltage due to the interaction between the current in feeders and the network impedance. This interaction may manifest as voltage drop due to the current drawn by loads or voltage rise due to the current injected by distributed generation. As such, voltage standards prescribe that network operators adopt a nominal operating voltage along with an allowable range. This allows for voltage variation within the network as well as efficient use of network assets whilst also limiting the magnitude of supply voltage variation to within the predetermined range. The process of maintaining voltage magnitudes within this allowable range is termed voltage regulation.

2.5.1 Voltage regulation without distributed generation

Prior to the deployment of significant volumes of distributed generation embedded into LV networks, LV feeders and the broader radial electricity distribution systems were designed for one-way power flow from central generators to the distribution transformer to end users. In this scenario, voltage levels are highest at the terminals of the distribution transformer (sending end) and decrease along the length of the LV feeder due to voltage drop. In such a scenario, the voltage will be lowest at the point farthest from the terminals of the distribution transformer.

A simplified representation of the way that voltage varies across a LV feeder under this scenario is illustrated in Fig. 2.8 where:

- V_0 is the magnitude of upstream voltage
- V_1 is the magnitude of the voltage at the distribution transformer LV terminals
- V_2 is the magnitude of the receiving end voltage
- I is the feeder current
- r and x represent the feeder resistance and reactance, respectively
- P_L and Q_L are the active and reactive load power, respectively

In this simple example, the magnitude of the voltage at the LV terminals of the distribution transformer (V_1) is dependent on the magnitude of the upstream voltage (V_0), the voltage drop across the distribution transformer

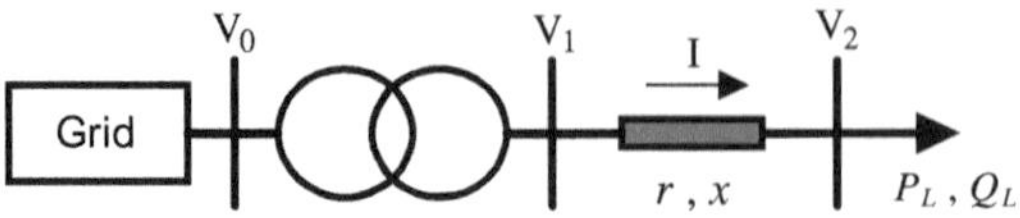

Fig. 2.8 Illustration of voltage regulation in an LV feeder without distributed generation.

and the distribution transformer turns ratio. The magnitude of the voltage at the receiving end of the feeder (V_2) is equal to the magnitude of the voltage at the distribution transformer terminals minus the voltage drop caused by the interaction between the feeder load current and the feeder impedance and can be expressed as follows:

$$\boldsymbol{V}_2 = \boldsymbol{V}_1 - \boldsymbol{I}(r + jx) \tag{2.1}$$

The voltage behaviour of the circuit depends on the power factor of the load. For the same load current and voltage, the phasor diagrams shown in Fig. 2.9 illustrate that, when compared with a purely resistive load, the supply voltage required to provide the specified load voltage is relatively greater for the inductive load case, whereas the supply voltage magnitude for a capacitive load may be less than the magnitude of the voltage at the load.

Voltage regulation can be defined as a measure of the variation of the voltage at a specific point in the power system in relation to the nominal level that is expected. Voltage regulation can be calculated using the definition

$$V_{\text{reg}} = \frac{(|V_S| - |V_L|)}{|V_L|} \times 100\% \tag{2.2}$$

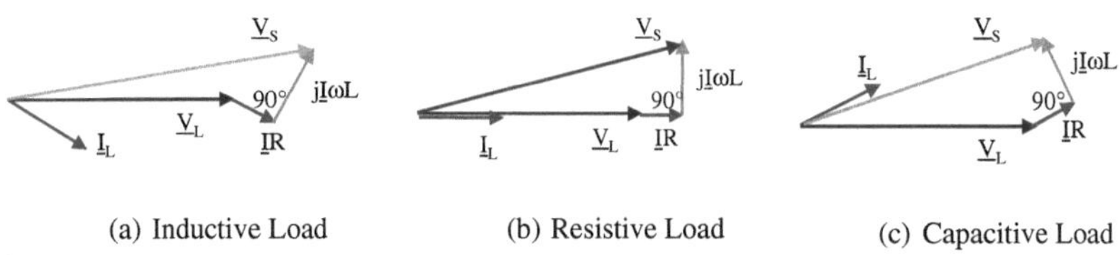

Where:

- I_L is the load current
- V_S is the source voltage (equivaluent to V_1 in Fig. 2.8)
- V_L is the voltage at the load (equivalent to V_2 in Fig. 2.8)
- IR is the voltage due to interaction between the load current and the resistance of the network
- jIωL is the voltage due to interaction between the load current and the reactance of the network

Fig. 2.9 Vector diagrams of voltage for different load power factors.

where

V_S is the sending end (source) voltage

V_L is the receiving end (load) voltage

By expressing Eq. (2.1) in terms of the power quantities (P and Q) it is possible to express voltage regulation in terms of the active and reactive power flowing in the circuit as follows:

- V_S is the source voltage
- V_L is the receiving end (load) voltage
- I is the current drawn by the load $= I\cos\theta + jI\sin\theta$
- $Z = R + jx$

Given $V_S = V_L + IZ$ (from Eq. 2.1) then

$$V_s = (V_L + IR\cos\theta - IX\sin\theta) + j(IX\cos\theta + IR\sin\theta) \tag{2.3}$$

If the voltage drops due to IR and IX are relatively small (as will be the case in the majority of LV networks), the quadrature component of Eq. (2.3) can be ignored. By substituting Eq. (2.3) into Eq. (2.2) and multiplying both numerator and denominator by V_R, a simplified voltage regulation can be expressed as follows

$$V_{\text{reg}} \approx \frac{r \cdot P_L + x \cdot Q_L}{V_L^2} \tag{2.4}$$

Eq. (2.4) shows that voltage regulation is a function of the active power flowing in the circuit multiplied by the resistance of the circuit and the reactive power flowing in the circuit multiplied by the reactance of the circuit. This has important implications for techniques to address voltage regulation issues as will be discussed further in Section 2.6.

As can be seen in Eq. (2.1), V_2 is dependent on the magnitude of the current (I) drawn by the loads. V_2 is also dependent on the impedance of the feeder which in turn is related to the distance of the receiving end of the feeder from the distribution transformer. This leads to the concepts of *light* (low) and *heavy* (high) loading. Periods of light and heavy loading are generally cyclic across hours of the day and days of the week. For residential LV networks, peak load generally occurs in the morning and early evening when residents are generally preparing breakfast and evening meals respectively. Periods of light load occur during the middle of the day and the late evening through to early morning. For LV networks supplying commercial loads, demand is generally highest during daylight hours and generally lower during night. Figs 2.10 and 2.11 show simple representations of the voltage profile along an LV feeder for light load and heavy load

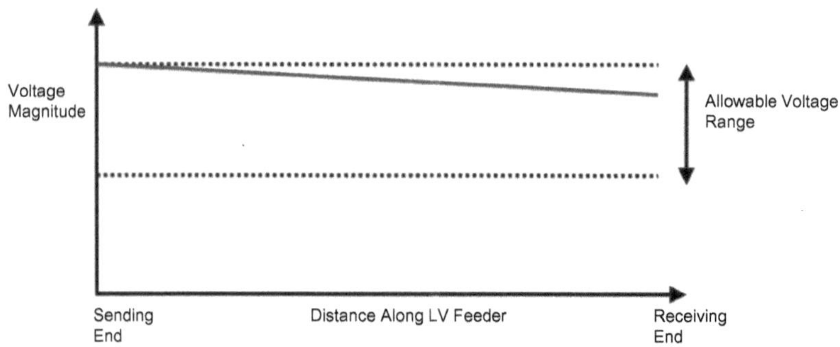

Fig. 2.10 LV feeder voltage profile under light load conditions.

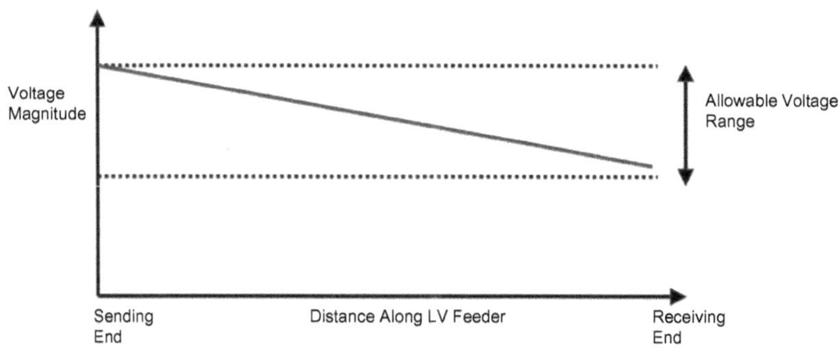

Fig. 2.11 LV feeder voltage profile under heavy load conditions.

conditions, respectively. Appreciation of the impacts of light and heavy loading on feeder voltage profiles is important as the network should be designed to ensure appropriate voltage magnitudes at the sending and receiving end of the feeder under all normal operating scenarios.

In LV networks, the distribution transformer plays an important role in voltage regulation. The standard (traditional) distribution transformer generally includes a number of buck and/or boost taps which can be manually adjusted but do not actively manage voltage regulation and cannot be operated/adjusted on-load. As such, for all intents and purposes, the distribution transformer has a fixed turns ratio. The transformer tap position is determined at the planning stage and is calculated based on the expected upstream voltage and voltage drop within the medium voltage network.

Fig. 2.12 shows photographs of example pole-mounted (Pole Top) and ground-mounted (Padmount) distribution transformers.

Fig. 2.13A shows a simplified representation of the schematic of the off-load tap changer in a distribution transformer. In this example, it can be seen that there are three tap positions: a neutral tap, a 5% boost tap and a 5% buck

(a) (b)

Fig. 2.12 Photographs of distribution transformers (A) Pole Top, (B) Padmount Substation (sometimes called Kiosk).

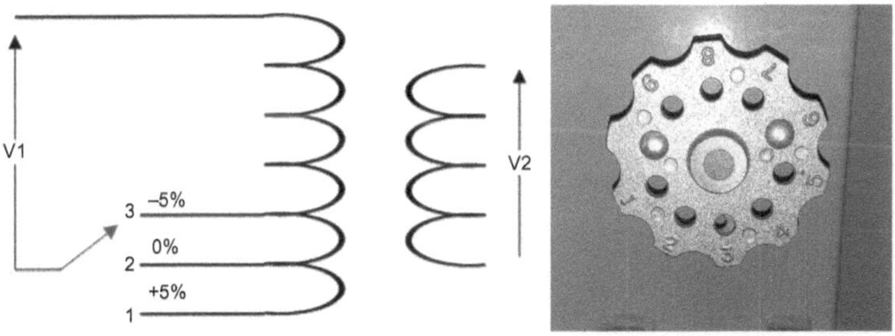

(a) (b)

Fig. 2.13 Off-load distribution transformer tap changers, (A) simple example schematic, (B) physical tap changer dial.

tap. Whilst this example only shows three taps, many distribution transformers will have up to 8 or 9 tap positions split giving a range of boost and buck options (a common boost/buck ration is 2.5%) as indicated by Fig. 2.13B which shows a photograph of the dial used to select the distribution transformer off-load tap position.

Fig. 2.14 shows a photograph of the internal construction of a distribution transformer which has been modified in order to demonstrate the working parts. The primary winding (MV) is on the far left-hand side of the core. The secondary (LV) winding can be seen in the centre, whilst the core only is on the far right.

Whilst standard distribution transformers do not have on-load tap changers, generally the upstream (medium voltage) is regulated by on-load tap changing transformers at the upstream substation (often termed the zone substation).

Fig. 2.14 Photograph of the internal construction of a distribution transformer.

2.5.2 Impact of distributed generation on voltage regulation

The penetration of distributed generation into LV networks has increased significantly over the past decade and continues to do so. The distributed generation technology that is predominantly being installed into LV networks is solar photovoltaic (PV) generators at the consumer premises. In many places these generators are referred to as rooftop solar. As an example of the proliferation of rooftop solar, Fig. 2.15 shows the installed capacity and average size of distributed solar PV installations in Australia (Australia has amongst the highest per capita rates of distributed solar PV penetration in the world). The figure clearly shows the incredibly rapid take-up of

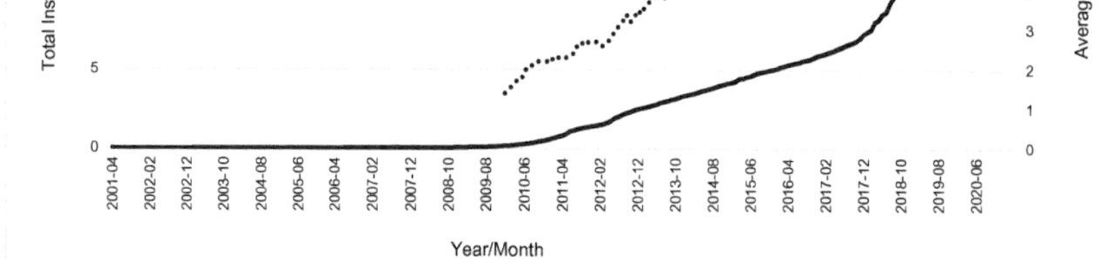

Fig. 2.15 Australian PV installations: monthly total capacity (MW) since 2001 and average installation size (kW) since 2010. *(Data sourced from Australian PV Institute (APVI), Solar Map, Funded by the Australian Renewable Energy Agency, Webpage (Last Accessed 7 April 2021).).*

dispersed generation in terms of both overall cumulative rating and increasing system size.

High levels of distributed generation can affect the operation of distribution networks and can cause serious problems in terms of voltage management, reverse power flows, overloading of feeders and transformers and the operation of voltage management control in the network. Arguably the most significant impact of increasing penetration of distributed generation into LV networks to date has been interference with existing voltage regulation practices.

2.5.2.1 Voltage regulation with distributed generation

Considering the performance of the LV network from an upstream perspective, local generation will decrease the load on LV feeders. In the first instance the distributed generation will supply local loads and voltage levels along the feeder will increase due to the decreased current being drawn from the upstream system, i.e. I from Eq. (2.1) reduces. In the event that the entire local demand is supplied by local, distributed generation, the voltage level at the receiving (far) end of the feeder will be equal to that of the LV terminals of the distribution transformer (this is analogous to the light load scenario presented in Fig. 2.10). If generation exceeds local demand, injection of current into the network will cause reverse power flow and will begin to cause the voltage along the feeder to rise (this may be thought of as reverse voltage drop).

Fig. 2.16 shows a simple illustration of voltage behaviour on an LV feeder with distributed generation present. In this case P_G and Q_G are the active and reactive power magnitudes of the distributed generation, respectively. In this scenario, generation at the end of the feeder will serve to reduce the voltage drop along the length of the feeder. Once local generation exceeds local load, power will flow from V_2 towards V_1, termed reverse power flow.

The amount of voltage rise is dependent on the inequality of local load compared to local generation as well as the impedance of the network. Voltage rise is a well-known phenomenon associated with distributed

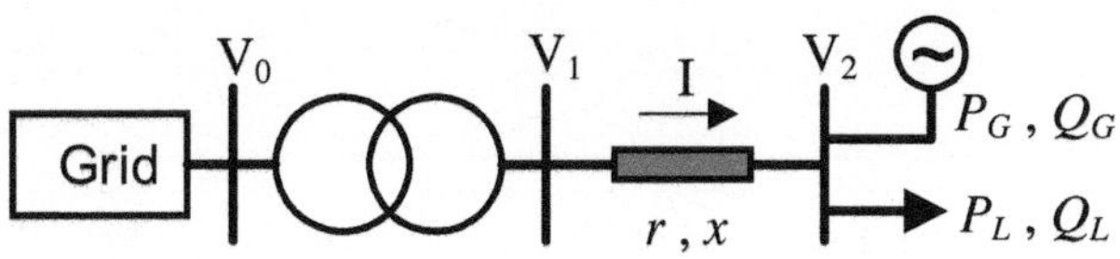

Fig. 2.16 Illustration of voltage regulation in an LV feeder without distributed generation.

generation. Until recently, the phenomena of voltage rise have been exacerbated by regulatory requirements for solar PV generators and feed-in tariff incentive schemes which required or made it desirable for inverters to inject only active power into the network. However, the power factor requirements for PV inverters are now being varied in many locations in order to assist with voltage regulation (this topic is discussed in detail in Section 2.6).

Figs 2.17 and 2.18 illustrate the impact of distributed generation on a feeder voltage profile for heavy load and light load scenarios, respectively. In this example, it can be seen that under light load, the distributed generation is capable of increasing the voltage to a magnitude above the allowable voltage range. Management of the voltage rise is one of the key challenges of integrating distributed generation into electricity supply systems.

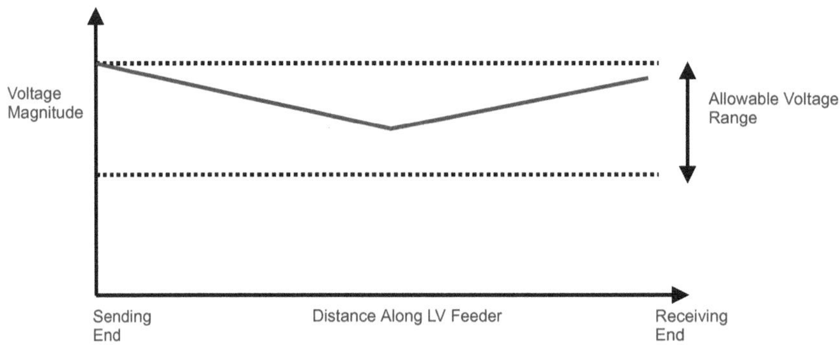

Fig. 2.17 LV feeder voltage regulation under heavy load conditions with distributed generation.

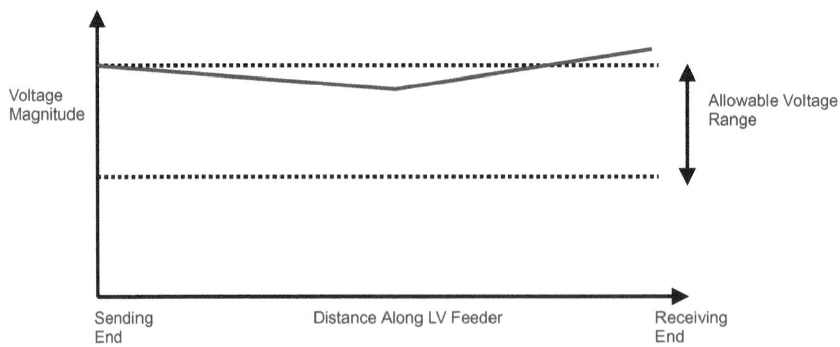

Fig. 2.18 LV feeder voltage regulation under light load conditions with distributed generation.

When distributed generation is installed in the LV network, the voltage equation takes the form of:

$$|\Delta V| \approx \frac{r(P_L - P_G) + x(Q_L - Q_G)}{|V_2|} \tag{2.5}$$

Appreciation of Eq. (2.5) is important in understanding how effective any voltage control strategies which aim to mitigate variation of voltage magnitudes through the adjustment of either active or reactive power (or power factor) will be.

2.6 Techniques for improving voltage regulation

This section discusses a number of the common techniques that are applied to mitigate long-term variations in voltage magnitude that exceed the allowable range.

2.6.1 Network upgrades

Throughout this chapter the relationship between management of voltage levels and network impedance has been evident. As such, one of the fundamental methods of better controlling voltage is to reduce systems impedance via networks upgrades. Network upgrades (i.e. reduction in impedance) can be effective in resolving both under and overvoltage problems and also increase the hosting capacity for distributed generation. Examples of networks upgrade include

- Reconductoring, i.e., replacement of conductor with one of a higher rating and lower impedance. Reconductoring is rarely used in practice as it is generally more expensive than most other options for addressing unacceptable voltage variation.
- Upgrading distribution transformers to a larger size (and lower impedance).
- Splitting feeders and installing additional distribution transformers.

Another very effective method of resolving voltage issues is to adjust the upstream voltage by varying the voltage regulation setting of the on-load tap changing transformer at the upstream (zone) substation. This can be achieved through static settings; however, more complex arrangements are beginning to be implemented. These include

- Bespoke voltage regulation droop curves which take into account solar generation profiles. In this scenario, the voltage is reduced during periods of high irradiance in order to account for the voltage rise due to distributed generation.

- Closed loop voltage control using the data from advanced metering infrastructure dispersed throughout the network to determine the set point for automatic voltage regulators.

2.6.2 Line capacitors

In LV networks without dispersed generation, the most significant issue for voltage regulation has traditionally been concerns related to undervoltage at the remote ends of the LV feeders under heavy load. In such situations, capacitors can be used to inject reactive power in order to increase voltage magnitudes. Fig. 2.9C provides an explanation of how capacitive current can increase voltage. The advantage of capacitors is that they are relatively cheap. However, the capacitors are generally constructed in discrete steps which are unable to provide voltage control across a continuous range. Further, capacitors themselves can be very sensitive and easily damaged by overvoltage, harmonic distortion and transients.

2.6.3 Power factor correction

Use of power factor correction in power systems is a common practice. Power factor correction is used both at plant level and at supply system level. The objective of power factor correction at a plant is generally driven by the requirements of the supply authorities. In contrast to domestic customers who, in most cases, pay only for real power consumption, large industrial customers (whose loads are generally inductive in nature) pay for both real and reactive power. Thus, having an improved power factor at the point of supply is an advantage for the plant and may be required in connection agreements. The most commonly applied technique to achieve a better power factor is the connection of a suitable capacitor bank at the plant. As such, whilst not specifically applied to address voltage concerns, capacitor banks used for power factor correction will operate in the same ways as line capacitors and can increase voltage levels (which may or may not be desirable).

2.6.4 Low voltage regulators

A low voltage regulator (LVR) increases (boost operation) or decreases (buck operation) the voltage magnitude within a band set around the nominal voltage. For example, a LVR may operate within a range of $\pm 13\%$ of the nominal voltage and can have a regulation accuracy of ± 1 V. In many cases

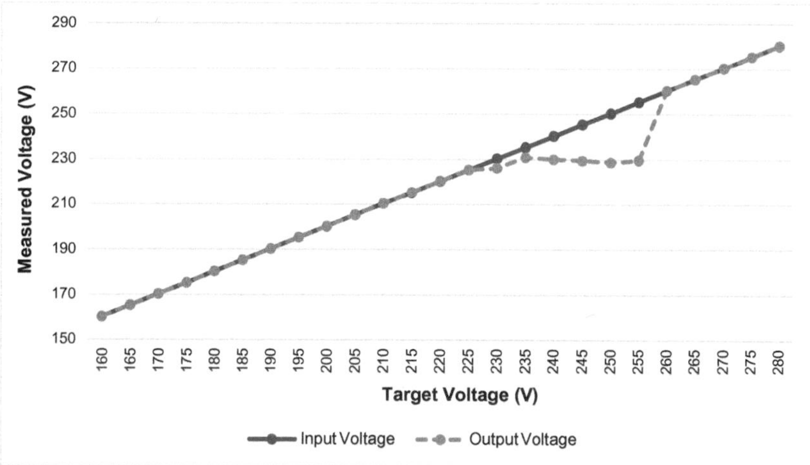

Fig. 2.19 Example of low voltage regulator operation.

the voltage regulator is designed to maintain a fixed output voltage (which can be defined by the user). An LVR with bidirectional capability is able to adjust the voltage when the power flow is reversed from the downstream network to the distribution substation. Fig. 2.19 shows an example of the voltage regulation capabilities for a LVR. In the case of this device, the target voltage was set to 230 V. It can be seen that the device is highly capable of regulating the output voltage to be close to the target voltage for input voltages between 225 and 255 V.

There are a number of ways to implement a LVR. A traditional LVR essentially operates as a small on-load tap changing transformer. The LVR includes a voltage sensing system that operates the tap changer to regulate the voltage magnitude. With the increase in distributed generation, more recent designs include a reversing polarity switch that sets the direction of voltage change. More recently, LVR topologies with increased complexity have become common, for example transformers with more complicated winding arrangements along with LVRs which use power electronic switching devices to regulate the voltage.

The main advantages of LVRs are the ability to install them at different locations and to regulate the voltage during reverse power flow. The major disadvantage is the cost, especially for low-density distribution networks. In general, voltage regulators used in LV networks are often of modest size (e.g. 10 kW) and as such may only be used for single installations.

2.6.5 Low voltage STATCOM

Static synchronous compensators (STATCOMs) have been applied in transmission systems for many years to regulate voltage and assist with stability. In simple terms, STACOMs work by injecting or absorbing reactive power in order to increase or decrease voltage magnitude.

In recent years, there has been a trend towards the application of STACOM technology in LV networks. Whilst smaller in rating, Distribution STATCOMs or D-STATCOMs use the same operating principles as those applied at higher voltage levels. The STACOM works by injecting reactive power according to a predefined droop curve. Fig. 2.20 shows an example of a droop curve for a LV STACOM.

A significant advantage of the LV STATCOM over passive devices such as switched capacitors and reactors is that it is capable of injecting or absorbing reactive power and as such can mitigate for both under and overvoltage scenarios. The STACOM also allows much finer control of the amount of reactive power being injected or absorbed. Another advantage of the STATCOM is that is installed in parallel to the supply network. As such, a failure of the STATCOM will not lead to a loss of supply which may be the case with series devices such as LVRs.

A potential disadvantage of the STATCOM is that its effectiveness is dependent on the X/R ratio of the distribution network; the STATCOM will be less effective in networks with low reactance. As such, the STACOM may not be a viable option in all scenarios.

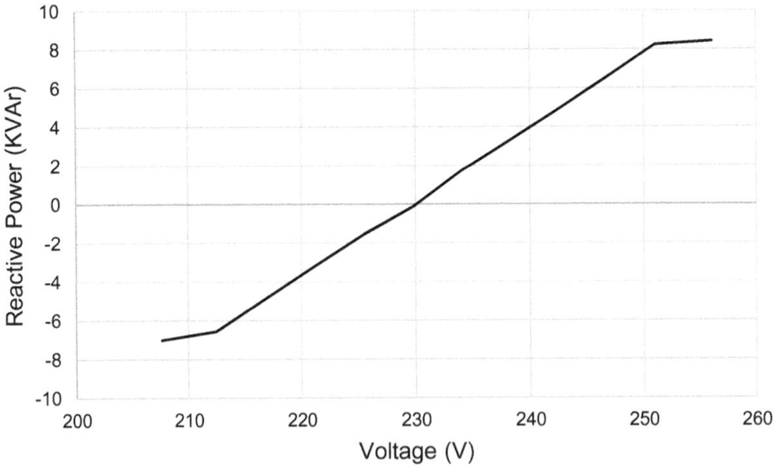

Fig. 2.20 Example LV STATCOM droop curve.

2.6.6 On-load tap changing distribution transformers

A LV on-load tap changing distribution transformer implements the same control philosophy as the on-load tap changing transformers at large substations. The transformer incorporates an automatic voltage regulating device which adjusts the tap position in order to maintain the voltage within a specified band. Whilst these types of transformers will regulate the voltage, their disadvantages include cost and maintenance requirements when compared to traditional distribution transformers with taps that must be adjusted manually off-load. Being a relatively new device, the service life of on-load tap changing transformers remains unknown when compared to the simple fixed ratio distribution transformers which are typically expected to be in service for up to 40 years.

2.6.7 Ferroresonant transformer

A ferroresonant transformer, also known as a constant voltage transformer (CVT), is a transformer that operates in the saturation region of the transformer B–H curve. Voltage sags down to 30% retained voltage can be mitigated through the use of ferroresonant transformers. Fig. 2.21 shows a schematic of a ferroresonant transformer.

The effect of operating the transformer in this region is that changes in input voltage only have a small impact on the output voltage. Ferroresonant transformers are simple and relatively maintenance free devices which can be very effective for small loads. Ferroresonant transformers are available in sizes up to approximately 25 kVA. A disadvantage of ferroresonant transformers is that they introduce extra losses into the circuit and are highly inefficient when lightly loaded and in some cases, may also introduce distorted voltages. In addition, unless greatly oversized, ferroresonant transformers are generally not suitable for loads with high inrush currents such as direct-on-line motors.

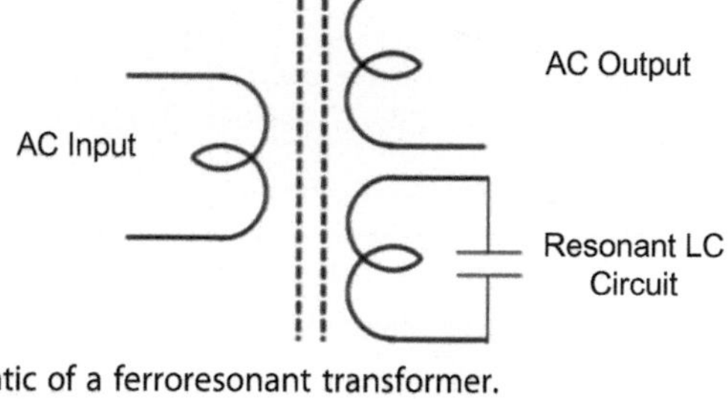

Fig. 2.21 Schematic of a ferroresonant transformer.

2.6.8 Inverter power quality response modes

The impact of distributed generation on voltage regulation, particularly voltage rise, is well understood. As distributed generation penetration has increased, focus has shifted to how inverters may be used to assist with voltage regulation (and increase hosting capacity). Many inverters whether they be for PV or batteries now incorporate, or are required by standards or legislation to incorporate, the capability to assist with voltage regulation. The in-built ability of inverters to dynamically vary performance in response to the voltage at the device terminals is termed power quality response modes. The two most widely applied power quality response modes are volt–watt response and volt–var response.

2.6.8.1 Volt–watt response

The volt–watt response mode varies the output active power depending on the voltage at the inverter terminals. As demonstrated by Eq. (2.4) the effectiveness of the volt–watt response is dependent on the magnitude of the resistance in the network. An example of a volt–watt response droop curve is shown in Fig. 2.22.

As can be seen, the inverter progressively lowers the active power (watt) output as the voltage exceeds the pickup voltage threshold resulting in the inverter injecting reduced active power as the voltage magnitude increases. The values of V_{w1} and V_{w2} are generally defined by the network operator. As an example, the power levels (with respect to rated power) may be 100% at V_{w1} and 20% at V_{w2} with V_{w1} being 253 V and V_{w2} being 260 V (in a 230-V nominal network).

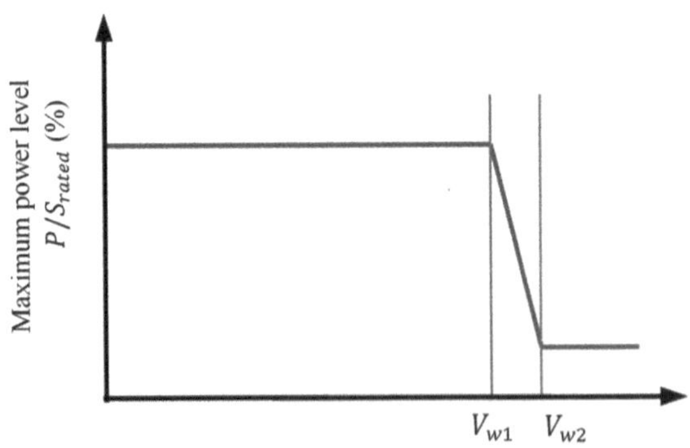

Fig. 2.22 Example volt–watt response droop curve.

2.6.8.2 *Volt–var response*

The volt–var response mode varies the inverter reactive power depending on the magnitude of the voltage at the inverter terminals. As demonstrated by Eq. (2.4) the effectiveness of the volt–var response is dependent on the magnitude of the reactance in the network. Fig. 2.23 shows an example of a volt–var response droop curve. The values V_1, V_2, V_3 and V_4 are generally specified by the network operator. As an example, default values as specified by the Australian standard for grid connected inverters (AS 4777) are as follows:

- V_1—207 V, 44% (injecting)
- V_2—220 V, 0%
- V_3—240 V, 0%
- V_4—258 V, 60% (absorbing)

As can be seen in Fig. 2.23, one of the advantages of the volt–var response mode is that it can be used to increase voltage when the magnitude is low as well as decrease the voltage magnitude when voltage magnitudes begin to get too high. However, it should be noted that some of the inverter rated capacity is required to inject/absorb the reactive power and as such, active power export may be limited during such scenarios.

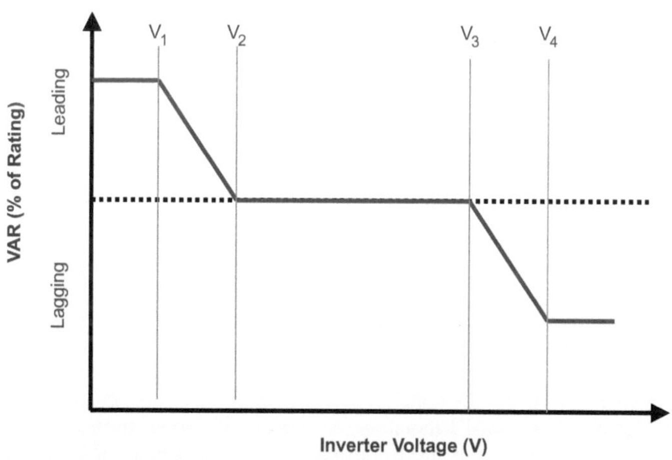

Fig. 2.23 Example volt–var response droop curve.

2.6.9 Energy storage

Local energy storage can be applied to assist with voltage regulation (specifically voltage rise) in the presence of high levels of distributed generation.

Energy storage may be used to absorb the active power injected by the local generation, reducing the amount exported into the supply network. This energy storage may take the form of batteries as well as alternate energy storage such as hot water. Whilst effective in theory, most studies indicate that the size of the energy storage compared to the size of the distributed generation is not sufficient to be able to store enough energy to provide an effective voltage regulation response—typically, the energy storage fills before peak generation (and peak voltage rise). Energy storage may present better opportunities to assist with voltage regulation as battery prices decrease. Another emerging opportunity related to energy storage is the possibility of using the on-board storage of electric vehicles to assist with management of voltage magnitudes.

2.6.10 Other technologies

There are a range of other technological solutions that can be used to mitigate voltage excursions/improve voltage regulation. In many cases, these solutions are utilised for mitigating voltage sags; however, they are equally capable of compensating for steady-state voltage regulation issues. Examples include uninterruptible power supplies and flywheel systems. Detailed explanations of the operating principles of these devices can be found in the chapter on voltage sags.

References

[1] J. David, P. Ciufo, S. Elphick, D. Robinson, Preliminary evaluation of the impact of sustained overvoltage on low voltage electronics-based equipment, Energies 15 (4) (2022) 1536.

[2] D. Lauria, Conservation voltage reduction (CVR) at northeast utilities, IEEE Trans. Power Deliv. 2 (4) (1987) 1186–1191.

[3] B. Scalley, D. Kasten, The effects of distribution voltage reduction on power and energy consumption, IEEE Trans. Educ. 24 (3) (1981) 210–216.

[4] The CSU, Chico Research Foundation, W.H. Wattenburg, Advanced switches for soft blackouts, critical infrastructure protection, unanticipated discovery of emergency voltage reduction for grid protection—PIER final project report, Report Prepared for California Energy Commission Public Interest Energy Research Program, 2006.

[5] Pacific Northwest National Laboratory, Evaluation of conservation voltage reduction (CVR) on a national level, Report Prepared for US Department of Energy, 2010.

[6] A. Pisani, The Validity of Conservation Voltage Reduction in Australian Distribution Networks to Reduce Both Consumer and Network Costs, University of Wollongong, Undergraduate Thesis Undergraduate, 2011.

CHAPTER 3

Impact and management of power system voltage unbalance

3.1 Introduction

Voltage unbalance in power systems, as introduced in Chapter 1, refers to a condition where the voltage magnitudes of the three phases (whether measured line to neutral or line to line) are not equal in magnitude and/or the angles between the corresponding phasors are not equal to 120 degrees.

There are a number of mechanisms that can lead to voltage unbalance in power systems, including uneven distribution of single-phase installations across the three phases (e.g. as in the case of public low voltage distribution networks), large single-phase installations such as line-to-line connected electric traction systems and electric arc furnaces (especially in the melting phase). Despite balancing of loads across the three phases, voltage unbalance can still arise as a result of unequal impedances of the different phases depending on the tower geometry and/or in the absence of transposition of the phase conductors. Power system open circuit and short circuit faults which are asymmetrical can also lead to voltage unbalance which is normally short term in nature as the protection systems will isolate such faults.

In the recant years, connection of single-phase roof top solar photo voltaic (PV) systems and increasing numbers of electric vehicle (EV) chargers to public low voltage distribution networks have become a concern from a voltage unbalance point of view in addition to other issues. The more large wind and solar farms which are remotely located and connected to transmission and subtransmission lines that are asymmetrical have become a concern to network operators although these generating systems inject balanced currents at their connection points.

Considering the currently evolving power systems at all voltage levels, due attention should be paid to managing the voltage unbalance so that such power systems functions securely ensuring safe and reliable operation of the connected equipment.

Applied Power Quality
https://doi.org/10.1016/B978-0-323-85467-2.00001-9

3.2 Commonly used definitions

Historically, the level of voltage unbalance has been quantified employing several definitions [1,2] but most international standards now employ sequence quantities as given in Eqs (3.1), (3.2) (also known as true definitions), in which the negative and zero sequence voltage unbalance levels as complex quantities are expressed as ratios of the positive sequence voltage.

$$\text{Negative sequence voltage unbalance factor} = \frac{V_2}{V_1} \tag{3.1}$$

$$\text{Zero sequence voltage unbalance factor} = \frac{V_0}{V_1} \tag{3.2}$$

where

$$V_1 = \text{positive sequence voltage} = \frac{1}{3}\left(V_a + aV_b + a^2V_c\right)$$

$$V_2 = \text{negative sequence voltage} = \frac{1}{3}\left(V_a + a^2V_b + aV_c\right)$$

$$V_0 = \text{negative sequence voltage} = \frac{1}{3}(V_a + V_b + V_c)$$

$$V_a, V_b, V_c = \text{fundemental line} - \text{to} - \text{neutral voltage phasors}$$

$$a = -\frac{1}{2} + j\frac{\sqrt{3}}{2}$$

In the case, where only the line-to-line voltage magnitudes V_{ab}, V_{bc} and V_{ca} are known, the absolute values $|V_1|$, $|V_2|$ (line–line) can be determined using the expressions given in Eqs (3.3), (3.4) [3]. In addition, if the line-to-neutral voltage magnitudes V_a and V_b are also known, the absolute value $|V_0|$ (line–neutral) can be determined using the expression given in Eq. (3.5) [3].

$$|V_1| = \sqrt{\frac{1}{12}\left[\left(\sqrt{3}V_{ab} + \sqrt{4V_{bc}^2 - \left(\frac{V_{bc}^2 - V_{ca}^2}{V_{ab}} + V_{ab}\right)^2}\right)^2 + \left(\frac{V_{bc}^2 - V_{ca}^2}{V_{ab}}\right)^2\right]} \tag{3.3}$$

$$|V_2| = \sqrt{\frac{1}{12}\left[\left(\sqrt{3}V_{ab} - \sqrt{4V_{bc}^2 - \left(\frac{V_{bc}^2 - V_{ca}^2}{V_{ab}} + V_{ab}\right)^2}\right)^2 + \left(\frac{V_{bc}^2 - V_{ca}^2}{V_{ab}}\right)^2\right]} \tag{3.4}$$

$$|V_0| = \frac{1}{6}\sqrt{\left[\frac{V_{bc}^2 - V_{ca}^2}{V_{ab}} - 3\frac{V_b^2 - V_a^2}{V_{ab}}\right]^2 + \left[\sqrt{4V_{bc}^2 - \left(V_{ab} + \frac{V_{bc}^2 - V_{ca}^2}{V_{ab}}\right)^2} - 3\sqrt{4V_b^2 - \left(V_{ab} + \frac{V_b^2 - V_a^2}{V_{ab}}\right)^2}\right]^2} \tag{3.5}$$

Magnitude of the negative sequence voltage factor as given in Eq. (3.1), which is commonly known as the voltage unbalance factor (VUF), can also be conveniently determined using the three line-to-line voltage magnitudes using Eqs (3.6), (3.7) [4].

$$\text{VUF} = \left|\frac{V_2}{V_1}\right| = \sqrt{\frac{1 - \sqrt{3 - 6\beta}}{1 + \sqrt{3 - 6\beta}}} \tag{3.6}$$

where

$$\beta = \frac{V_{ab}^4 + V_{bc}^4 + V_{ca}^4}{\left(V_{ab}^2 + V_{bc}^2 + V_{ca}^2\right)^2} \tag{3.7}$$

3.3 Measurement of voltage unbalance

In the measurement of voltage unbalance as per IEC61000-4-30 [5], the fundamental frequency component of the line voltages will be used to calculate the voltage unbalance averaged over 3-s, 10-min and 2-h windows, where the basic measurement window is 10 cycles in 50-Hz systems. For example, the 3-s voltage unbalance factor is given as in Eq. (3.8).

$$\text{VUF}_{3\,\text{s}} = \sqrt{\frac{\left(\sum_{k=1}^{15} \text{VUF}_k^2\right)}{15}} \tag{3.8}$$

where

VUF_k= voltage unbalance factor determined over a 10-cycle window. Consecutive $VUF_{3\ s}$ values will be used to determine $VUF_{10\ min}$ and the consecutive $VUF_{10\ min}$ values will be used to calculate $VUF_{2\ h}$.

3.4 Impact of voltage unbalance

As is well known, the primary concern with regard to excessive voltage unbalance is the impact on mains connected three-phase rotating machines (both asynchronous and synchronous machines) where it can lead to vibration and noise, and hence can cause damage to structural components, increased heating in the windings leading to their premature failure.

Behaviour of three-phase induction machines subjected to voltage unbalance is well covered in the literature. Some key performance characteristics of a 4-kV, 200-kW three-phase induction motor operating at near rated load (P_mot—real power, Q_mot—reactive power) delivering a near constant mechanical torque (T_m) subjected to a small voltage unbalance are illustrated in the EMT [6] simulation outcomes in Fig. 3.1. The fluctuations in the electromagnetic torque (T_e) and the speed (W) of the machine are naturally associated with the noise and vibration where such levels will primarily depend on the level of voltage unbalance. Noting the fact that voltage unbalance appearing on the terminals of a three-phase induction motors establishes a reverse rotating magnetic field (caused by the negative sequence voltages) will lead to a rotor slip of $2\text{-}s_1$ where s_1 is the rotor slip associated with the forward rotating magnetic field (caused by the positive sequence voltages). Noting the fact that at near rated speed s_1 is very small, the magnitude of the positive sequence impedance Z_1 of the machine is relatively large compared to the magnitude of the negative sequence impedance Z_2, and $|Z_1/Z_2|$ ($259\,\Omega/36\,\Omega = 7.2$) approximately equal to the ratio of the starting current to running current $\frac{I_{\text{starting}}}{I_{\text{running}}}$.

Noting the relatively small Z_2, even a small negative sequence voltage will lead to a large current which leads to excessive stator and rotor heating and hence induction motors need to be derated under terminal voltage unbalance conditions. For example, as described in Ref. [7] according to the derating curve shown in Fig. 3.2, with 5% voltage unbalance, motors in general have to be derated to 75% of the load whereas they should be designed to handle 1% voltage unbalance without derating.

Whilst mains connected three-phase induction motors are seen to be the most affected by voltage unbalance, synchronous generators, especially the

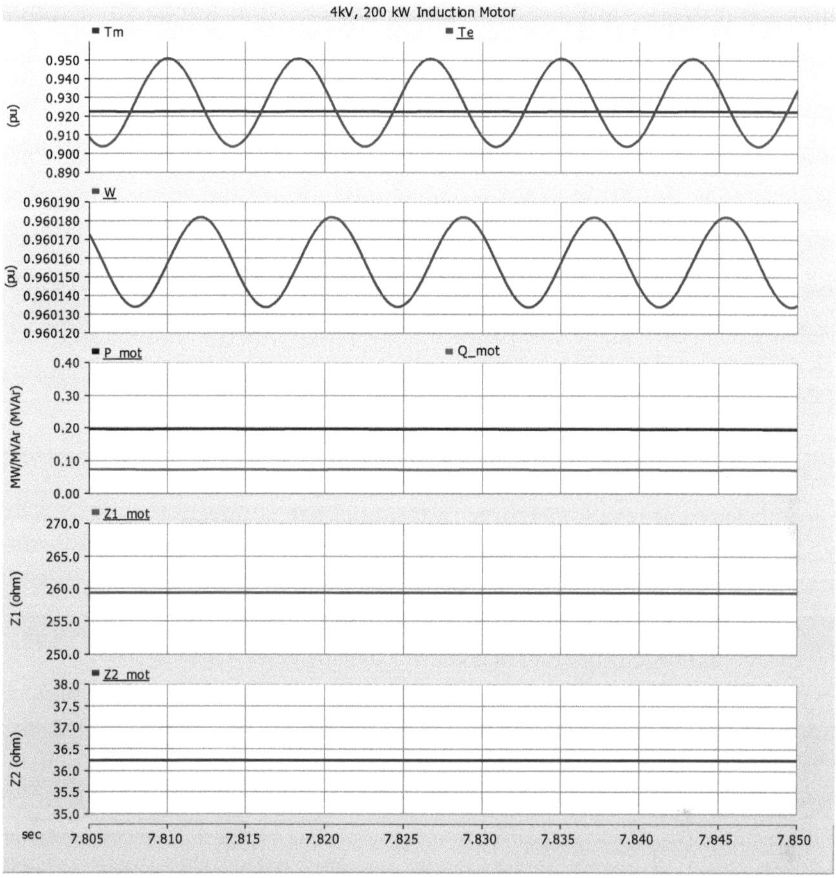

Fig. 3.1 Performance characteristics of a 4-kV, 200-kW three-phase induction machine operated with a VUF ≅0.35%.

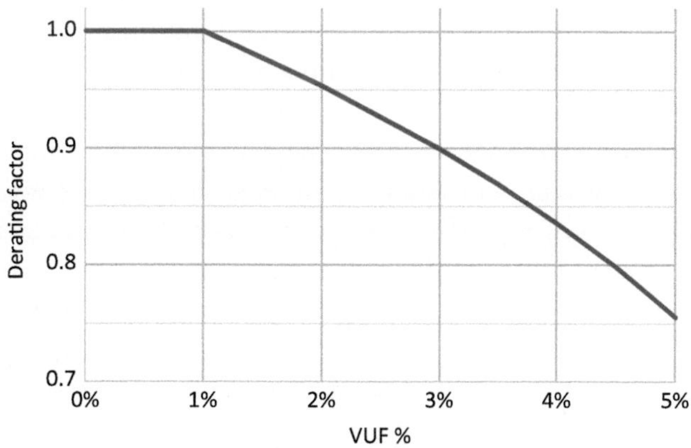

Fig. 3.2 Three-phase induction motor derating curve.

turbo alternators, also need to be protected from excessive negative sequence currents and hence the negative sequence voltages.

In addition, in the presence of supply voltage unbalance, mains connected three-phase power electronic converters can lead to generation of uncharacteristic harmonic currents which can contribute to additional voltage harmonic distortion in the ac supply [8].

3.5 Management of voltage unbalance in power systems

Management of voltage unbalance in power systems is a multifaceted task where due consideration has to be paid to a range of interrelated aspects which include

- causes of voltage unbalance
- limitation of unbalanced current emission
- fault levels at locations where installations are connected
- network compatibility levels
- transmission and distribution network design and planning practices
- applicable standards covering voltage unbalance limits
- network modelling and analysis

As per the International Electrotechnical Commission (IEC) philosophy, to ensure electromagnetic compatibility in networks, the voltage unbalance compatibility limits are specified for both low voltage and medium voltage networks [9,10] and the aim is to limit the voltage unbalance levels to be within the recommended planning limits at network level [4] to ensure that they do not exceed the compatibility levels. With this approach it is expected that the connected equipment, which are designed taking in to account voltage unbalance compatibility levels, will operate as expected.

Three-phase equipment that are connected at locations where the system impedance levels are relatively high (low fault levels) naturally can lead to voltage unbalance if the connected equipment are unbalanced. The system design needs to cater for some unbalance in the loads whilst aiming to limit the resulting voltage unbalance levels. The network planning needs to take into account the load growth and the technology changes (e.g. electric vehicle charging and roof top solar photovoltaic system) which can introduce voltage unbalance problems leading to operational issues.

Public low voltage (LV) network voltage unbalance is seen to be more problematic as there are a vast number of single and balanced and unbalanced three-phase loads which operate in a highly variable manner throughout the day. The management of voltage unbalance in such networks is seen to be the network owner responsibility using appropriate network design and operational guidelines, and it is sometimes required to ensure that customer equipment current unbalance emission levels are prescribed to ensure non-exceedance of adopted planning levels [11].

With regard to existing medium voltage (MV), high voltage (HV) and extra high voltage (EHV) networks, the management of voltage unbalance is usually guided by applicable rules or standards in the connection of installations. Complex unbalance network analysis software can be utilised in relevant studies where the data input requirements are demanding such as unbalanced load details and their operational regimes, network details consisting of tower geometries as some line geometries can play a significant role in network voltage unbalance despite the fact that the loads are balanced. In the application of standards and software tools in the management of voltage unbalance, greater insights need to be developed in relation to the relevant methodologies and the outcomes. Often these outcomes can be examined employing simpler networks rather than large complex networks in the first instance.

3.5.1 Installations supplied by a dedicated symmetrical network—Approximate analysis

Unbalanced currents drawn by installations are a primary cause of unbalanced voltages. Where this is the case, the various sequence domain terms associated with voltage unbalance that arise at the point of connection of an unbalanced installation supplied by a network assist with the voltage unbalance management process. As an explanation, consider the block diagram of the system shown in Fig. 3.3 where the source supplying the installation is assumed to be balanced with its positive sequence voltage, V_{1s}.

The sequence components of the voltage (V_0, V_1, V_2) at the point of evaluation (POE) are given in Eq. (3.9),

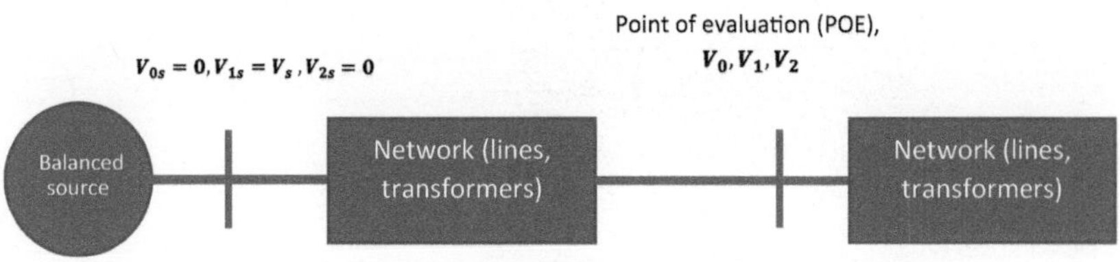

Fig. 3.3 An installation supplied by a balanced source through a network.

$$\begin{bmatrix} -V_0 \\ V_{1s} - V_1 \\ -V_2 \end{bmatrix} = \begin{bmatrix} Z_{00} & Z_{01} & Z_{02} \\ Z_{10} & Z_{11} & Z_{12} \\ Z_{20} & Z_{21} & Z_{22} \end{bmatrix} \begin{bmatrix} I_0 \\ I_1 \\ I_2 \end{bmatrix} \tag{3.9}$$

For a symmetrical network, as the off-diagonal elements: $Z_{01}=Z_{10}$, $Z_{02}=Z_{20}$, $Z_{12}=Z_{21}$ are all equal to zero, the respective sequence voltages are relatively easy to be establish. For example, $V_2=-Z_{22}I_2$ can be used to establish the magnitude of VUF as given in Eq. (3.1), and by making the assumption that the magnitude of the positive sequence voltage $|V_1|$ is approximately equal to the nominal value of the line-to-neutral voltage V_{Ln}, and by noting for passive networks $Z_{22}=Z_{11}$,

$$\text{VUF} = \left|\frac{V_2}{V_1}\right| = \frac{|-I_2 Z_{11}|}{V_{Ln}} \tag{3.10}$$

Further, noting that impedance Z_{11} is inversely proportional to the three-phase fault level or the short circuit capacity S_{sc}, Eq. (3.10) can be re-expressed as follows:

$$\text{VUF} = \frac{3V_{Ln}}{S_{sc}}|I_2| \tag{3.11}$$

which can be re-written substituting $I_2 = \frac{1}{3}\left(I_a + a^2 I_b + a I_c\right)$

$$\text{VUF} = \frac{V_{Ln}}{S_{sc}}\left|\left(I_a + a^2 I_b + a I_c\right)\right| \tag{3.12}$$

For a constant power installation, supplied by a 4-wire system (i.e. star connected installation with a neutral), the three line-to-neutral complex powers $S_a=P_a+jQ_a$, $S_b=P_b+jQ_b$, $S_c = P_c+jQ_c$ can be used to write approximate expressions for I_a, I_b and I_c, using phase '*a–n*' voltage as the reference phase with a magnitude V_{Ln},

$$I_a = \frac{S_a^*}{V_{Ln}} \tag{3.13}$$

$$I_b = \frac{S_b^*}{aV_{Ln}} \tag{3.14}$$

$$I_c = \frac{S_c^*}{a^2 V_{Ln}} \tag{3.15}$$

Substituting Eqs (3.13)–(3.15) in Eq. (3.12), an expression for VUF can be established as [12] follows:

$$\mathrm{VUF} = \frac{\left|\left(S_a + a^2 S_b + aS_c\right)\right|}{S_{sc}} \tag{3.16}$$

For an installation supplied by a 3-wire system (i.e. delta connected system), using the three line-to-line complex powers $S_{ab}=P_{ab}+jQ_{ab}$, $S_{bc}=P_{bc}+jQ_{bc}$, $S_{ca}=P_{ca}+jQ_{ca}$, Eq. (3.16) can be re-expressed as [12] follows:

$$\mathrm{VUF} = \frac{\left|\left(a^2+1\right)S_{ab} - S_{bc} + (a+1)S_{ca}\right)\right|}{S_{sc}} \tag{3.17}$$

For an installation that consists of both line-to-neutral and line-to-line connected loads, Eqs (3.16), (3.17) can be combined to give [12],

$$\mathrm{VUF} = \frac{\left|\left(S_{p1} + S_{p2} + S_{p3}\right)\right|}{S_{sc}} \tag{3.18}$$

where

$$S_{p1} = \sum S_a + \left(a^2+1\right)\sum S_{ab}$$

$$S_{p2} = a^2\sum S_b - \sum S_{bc}$$

$$S_{p3} = a\sum S_c + (a+1)\sum S_{ca}$$

Example: Consider the case of a 200-kVA, 11-kV/400-V three-phase transformer having an impedance of 4% based on its own rating supplying a constant power 50-kVA installation at unity power factor. Assume that the 11-kV side of the transformer is supplied by an ideal source.

Calculate the VUF at the POE, if the installation is connected (a) line-to-neutral (*a–n*) and (b) line-to-line (*a–b*).

The three-phase short circuit capacity at POE,

$$S_{sc} = \frac{200\mathrm{kVA}}{4\%} = 5000\ \mathrm{kVA}$$

(a) Using Eq. (3.16), $\mathrm{VUF} = \frac{\left|\left(S_a + a^2 S_b + aS_c\right)\right|}{S_{sc}} = \frac{|S_a|}{S_{sc}} = \frac{|50+j0|}{5000} = 1\%$

(b) Using Eq. (3.17), $\mathrm{VUF} = \frac{\left|\left(a^2+1\right)S_{ab} - S_{bc} + (a+1)S_{ca}\right)\right|}{S_{sc}} = \frac{\left|\left(a^2+1\right)S_{ab}\right)\right|}{S_{sc}}$

$$= \frac{\left|\left(a^2+1\right)(50+j0)\right|}{5000} = 1\%$$

3.5.2 Symmetrical installations supplied by a dedicated asymmetrical network—Approximate analysis

For an asymmetrical network (e.g. an untransposed transmission line), the off-diagonal elements, $Z_{01}=Z_{10}$, $Z_{02}=Z_{20}$, $Z_{12}=Z_{21}$, are nonzero and hence the sequence voltages at the POE will demonstrate the effect of cross-coupling between sequence networks. For example, the negative sequence voltage at the POE given in Eq. (3.9) will be governed not only by the negative sequence current of the installation but also will depend on the zero and positive sequence currents drawn by the installation and the associated sequence cross-coupling impedances,

$$V_2 = -(Z_{20}I_0 + Z_{21}I_1 + Z_{22}I_2) \tag{3.19}$$

In the absence of zero sequence current (i.e. $I_0=0$) while noting the fact that $Z_{22}=Z_{11}$for passive networks, Eq. (3.19) can be written as follows:

$$V_2 = -(Z_{21}I_1 + Z_{11}I_2) \tag{3.20}$$

It is to be noted that even if the installation is completely balanced (i.e. $I_2=0$) there will be some negative sequence voltage that will arise at the POE, given by the first term $Z_{21}I_1$ of Eq. (3.20). The significance of this term will depend on the degree of asymmetry of the network which in fact is related to tower/conductor configurations as illustrated in Fig. 3.4 [13] from which it is clearly visible how the symmetry of the phase conductor configuration impact on the magnitude of Z_{12} compared to the magnitude of Z_{11}.

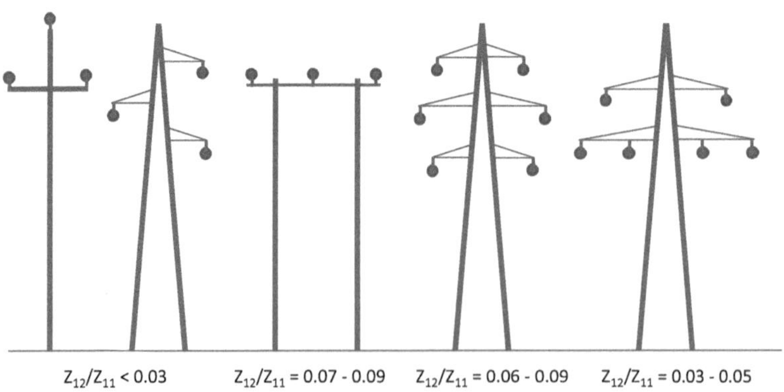

Fig. 3.4 Different tower configurations and indicative values of $\left|\frac{Z_{12}}{Z_{11}}\right|$ ratios.

Having a smaller $\left|\frac{Z_{12}}{Z_{11}}\right|$ ratio is an advantage with long lines. At the same time, even a short line with a relatively large $\left|\frac{Z_{12}}{Z_{11}}\right|$ ratio can become problematic where this is demonstrated by the example given as follows based on real world data [14].

Example: Different untransposed lines of different lengths in a 66-kV system supplying balanced unity power factor constant power loads are considered as follows. Assuming the supply source is balanced calculate the magnitude of the negative sequence voltage and the VUF at the end of the line assuming zero sequence current to be absent.

(a) 67.65 km, 22.86 MVA. $Z_{11} = 24.09\angle 73^{\circ}\,\Omega$,

$$Z_{12} = 0.62\angle 84^{\circ}\,\Omega \left(\left|\frac{Z_{12}}{Z_{11}}\right| = 0.03\right).$$

(b) 45.37 km, 17.15 MVA, $Z_{11} = 22.81\angle 45^{\circ}\,\Omega$,

$$Z_{12} = 1.25\angle 30^{\circ}\,\Omega \left(\left|\frac{Z_{12}}{Z_{11}}\right| = 0.05\right).$$

(c) 11.14 km, 28.58 MVA, $Z_{11} = 4.01\angle 72^{\circ}\,\Omega$,

$$Z_{12} = 0.31\angle 148^{\circ}\,\Omega \left(\left|\frac{Z_{12}}{Z_{11}}\right| = 0.08\right).$$

Case (a):

The positive sequence current is approximately given by:

$$|I_1| = \frac{22.86}{\sqrt{3}x66}\ \text{kA} = 200\ \text{A}, |I_2| = 0$$

$$|V_2| = |-(Z_{21}I_1 + Z_{11}I_2)| = |0.62\angle 84^{\circ} x200| = 124\ \text{V}$$

$$\text{VUF} = \frac{0.124}{66/\sqrt{3}} = 0.33\%$$

Repeating the same process,

Case (b):

$$|I_1| = \frac{17.15}{\sqrt{3}x66}\ \text{kA} = 150\ \text{A}$$

$$|V_2| = |1.25\angle 30^{\circ} x150| = 187.5\ \text{V}$$

$$\text{VUF} = \frac{0.1875}{66/\sqrt{3}} = 0.49\%$$

Case (c):

$$|I_1| = \frac{28.58}{\sqrt{3}x66}\ \text{kA} = 250\ \text{A},$$

$$|V_2| = |0.31\angle 148^{\circ} x250| = 77.5\,\text{V}$$
$$\text{VUF} = \frac{0.0775}{66/\sqrt{3}} = 0.2\%$$

As demonstrated by these examples, although $|I_1 Z_{21}|$ is used to evaluate the negative sequence voltage unbalance at the POE (also used in Ref. [4]), further modifications are required with regard to different types of loads as they behave differently when supplied through asymmetrical networks. To establish the negative sequence voltage unbalance factor at the POE with regard to constant power, constant impedance and constant current loads, the approximate expressions [14] which utilise the voltage regulation VR_t $\left(= \left|\frac{I_1 Z_{11}}{V_{Ln}}\right|\right)$ of the line can be used which have been shown to be in close agreement with the corresponding unbalanced load flow outcomes compared to the use of $|I_1 Z_{21}|$.

Constant power loads:

$$\text{VUF} = \left|\frac{V_2}{V_{Ln}}\right| \approx \frac{1}{V_{Ln}}|-(Z_{21}I_1)|(1 - VR_t)^{\beta} \tag{3.21}$$

$\beta \approx -1$ for low power factor loads (~ 0.9)
≈ -2 for high power factor loads (~ 1.0)

Constant impedance loads:

$$\text{VUF} = \left|\frac{V_2}{V_{Ln}}\right| \approx \frac{1}{V_{Ln}}|-(Z_{21}I_1)|(1 - VR_t) \tag{3.22}$$

Constant current loads:

$$\text{VUF} = \left|\frac{V_2}{V_{Ln}}\right| \approx \frac{1}{V_{Ln}}|-(Z_{21}I_1)| \tag{3.23}$$

Induction motor loads:

Compared to three passive loads which are characterised by similar positive and negative sequence impedances, induction motors offer distinctly different positive and negative sequence impedances. For this reason, the evaluation of VUF at the POE of a bus bar which supplies induction motor loads that are usually connected at downstream low voltage bus bars in addition to other loads requires a different treatment as evident from the expression given in Eq. (3.24).

$$\mathrm{VUF} = \left|\frac{V_2}{V_{Ln}}\right| \approx \frac{1}{V_{Ln}}|-(Z_{21}I_1)|\frac{1}{\left[1+\left(\frac{VR_t}{1-VR_t}\right)\left(\frac{1}{\frac{1}{k_s}+\frac{1}{k_p}}\right)\right]} \quad (3.24)$$

where

k_s = ratio between the positive and negative sequence impedances of the aggregated motor loads supplied by the aggregated low voltage downstream bus bar (typical range 5–7)

$$k_p = \frac{S_{sclv}}{S_{mlv}}$$

where

S_{mlv} = total load supplied by the downstream low voltage bus bar (in MVA)

S_{sclv} = short circuit capacity of the aggregated low voltage bus bar (in MVA) ignoring the contribution from the network upstream of the POE

3.5.3 Asymmetrical Installations supplied by a dedicated asymmetrical network—Approximate analysis

So far, the cases considered assumed that the installations supplied are balanced; however, this may not be the situation practically. In the case where both the supply network and installations are asymmetrical, as per Eq. (3.19), both terms $Z_{21}I_1$ and $Z_{11}I_2$ require evaluation to establish the net V_2 at the POE.

For the system shown in Fig. 3.3, for the general case, VUF at the POE can be expressed as follows:

$$\mathrm{VUF} = \left|\frac{V_2}{V_{Ln}}\right| = \frac{1}{V_{Ln}}|-(Z_{20}I_0 + Z_{21}I_1 + Z_{22}I_2)| \quad (3.25)$$

For a constant power installation (as in the case of derivation of Eq. (3.16)) by using the nominal voltage at the POE, the approximate expression given in Eq. (3.26) can be derived (noting $Z_{22} = Z_{11}$),

$$\mathrm{VUF} = \left|\frac{V_2}{V_{Ln}}\right| = \frac{1}{V_{LL}^2}\left|Z_{20}^*\left(S_a + a\,S_b + a^2 S_c\right) + Z_{21}^*\left(S_a + S_b + S_c\right)\right.$$
$$\left. + Z_{11}^*\left(S_a + a^2 S_b + aS_c\right)\right| \quad (3.26)$$

Example

An untransposed, 12.47-kV, 3.2187-km line (1.143-m horizontal phase conductor separation) is characterised by the following sequence impedance matrix (Ω) (based on a test networks used in Refs. [14–16]:

$$\begin{bmatrix} Z_{00} & Z_{01} & Z_{02} \\ Z_{10} & Z_{11} & Z_{12} \\ Z_{20} & Z_{21} & Z_{22} \end{bmatrix} = \begin{bmatrix} 1.1838+5.9125i & 0.0486-0.0281i & -0.0486-0.0281i \\ -0.0486-0.0281i & 0.6122+1.2673i & -0.0972+0.0561i \\ 0.0486-0.0281i & 0.0972+0.0561i & 0.6122+1.2673i \end{bmatrix}$$

A three-phase installation at the receiving end of the line supplies three single-phase 3.33-MVA loads at power factors 0.85, 0.9 and 0.5, respectively, across the a, b and c phases (all lagging).

Using Eq. (3.26), first term gives (corresponding to $Z_{20}I_0$),

$$\frac{1}{V_{LL}^2}\left[Z_{20}^*\left(S_a + a\,S_b + a^2S_c\right)\right] = 0.0683\angle 53.04°\%$$

from the second term (corresponding to $Z_{21}I_1$)

$$\frac{1}{V_{LL}^2}\left[Z_{21}^*\left(S_a + S_b + S_c\right)\right] = 0.6975\angle 9.10°\%$$

from the third term (corresponding to $Z_{22}I_2$)

$$\frac{1}{V_{LL}^2}\left[Z_{11}^*\left(S_a + a^2S_b + aS_c\right)\right] = 1.5705\angle - 179.52°\%$$

giving a net VUF = 0.85 % at the POE.

It can be noted the transmission line has a relatively large $\left|\frac{Z_{12}}{Z_{11}}\right|$ ratio equal to approximately 0.08, and hence the line asymmetry is seen to introduce a significant negative sequence voltage unbalance of 0.6975%. However, comparing its phase angle (9.10°) with that caused by the asymmetry of the load (−179.52°), the net VUF at the POE is only 0.85% as illustrated by the phasor diagram of Fig. 3.5 (ignoring the relatively small contribution associated with $Z_{20}I_0$) taking the VUF at the POE as the reference.

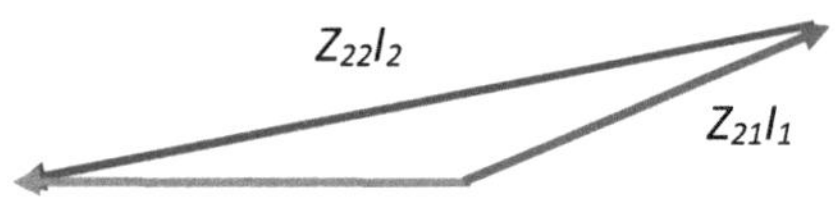

Fig. 3.5 Illustration of components of negative sequence voltage unbalance at the POE of the installation.

If the line asymmetry was completely eliminated, the VUF at the POE will be caused only by the installation asymmetry and will be equal to 1.5705% as the sequence impedance matrix becomes simply diagonal (with no changes to the diagonal elements).

If the same 3.33-MVA loads were differently distributed across the three phases with regard to their power factors, the outcomes with the net VUF at the POE would be different if the line is asymmetrical whilst it will be equal to 1.57% if it is symmetrical. These results summarised in Table 3.1 (shown within brackets are the VUF values obtained using network simulation) indicate the effect of the phase angles associated with the three terms in Eq. (3.19) (encompassing Z_{20}, Z_{21} and Z_{22}) although the individual magnitudes of the three terms (0.0683, 0.6975 and 1.5705, respectively) are independent of the earlier power factor distribution (0.85, 0.9 and 0.5, respectively). The agreement between the mathematical formulation and network simulation is seen to be reasonable for practical purposes although accurate formulations are available in Ref. [15] covering a number of load types.

3.5.4 Meshed asymmetrical network supplying asymmetrical loads—Approximate analysis

Noting the complexity of the voltage unbalance that arise as a result of the asymmetry of a single line supplying an isolated installation, in a meshed network, the determination of voltage unbalance levels at the various bus bars and their management is a complex task. Noting that untransposed lines can play a key role, the actual line geometries or the associated phase impedance matrices of all lines need to be considered at the data input stages in

Table 3.1 Summary of negative sequence voltage unbalance factor values at the point of evaluation.

Phase 'a' load power factor	0.85	0.5	0.9
Phase 'b' load power factor	0.9	0.85	0.5
Phase 'c' load power factor	0.5	0.9	0.85
VUF with asymmetrical line	0.85% (0.86%)	2.03% (1.82%)	1.99% (1.42%)
VUF with symmetrical line	1.57% (1.22%)	1.57% (1.22%)	1.57% (1.22%)

unbalanced load flow programmes that can be used to analyse voltage unbalance in such networks. At the same time, any known load asymmetries also need to be taken into account. Although unbalanced load flow programmes can be used in the study of voltage unbalance in asymmetrical networks, the outputs of such programmes do not necessarily explicitly provide an insight on the various contributors (e.g. installations and lines) to voltage unbalance at bus bar levels. Further, they do not provide guidance on systematic voltage unbalance mitigation approaches in problematic situations although trial and error approaches can be applied by synthetically transposing the lines and balancing the loads.

Based on detailed studies undertaken on a subtransmission network [14] with regard to meshed three-wire networks, an approximate but an elegant approach developed introduces the concept of the 'global emission vector', thus allowing systematic insights to be developed with regard to voltage unbalance. These global emission vectors characterise the contribution of every line and every load to the VUF in a network wide sense.

In applying the methodology using unbalanced load flow analysis, firstly, all loads and lines are synthetically balanced, except a single line. It is hypothesised that network wide voltage unbalance arises as a result of that single untransposed (unbalanced) line associated with its $Z_{21}I_1$. Although various bus bars exhibit varying magnitudes of negative voltage unbalance levels as a result of the considered untransposed line, they are characterised by a unique near constant phase angle associated with the term $Z_{21}I_1$. This product is identified as the global emission vector of the line and depending on the number of lines in a meshed network, similar number of unique global emission vectors would exist where those vectors will collectively be responsible for the network wide voltage unbalance when all lines are in operation.

As an example, for a network with 5 lines, the individual global emission vectors (labelled 1–5 with different magnitudes and phase angles) are placed in a polar plot together with their resultant (labelled R-lines with its magnitude and phase angle) illustrated in Fig. 3.6. Whilst the voltage unbalance emission capacity of the individual lines is represented by the magnitude of each vector, their placement with respect to the resultant global emission vector indicates the influence made on a network wide sense. The vector plot visually indicates that lines 2, 3 and 4 have a positive influence (leading to an increase in unbalance), whereas line 1 does not make any influence and line 5 has a negative influence (leading to a decrease in unbalance), thus allowing the identification of those dominant contributors to voltage unbalance in a network wide sense. The vector plot also indicates a prioritised approach for reduction of voltage unbalance on a network wide sense where

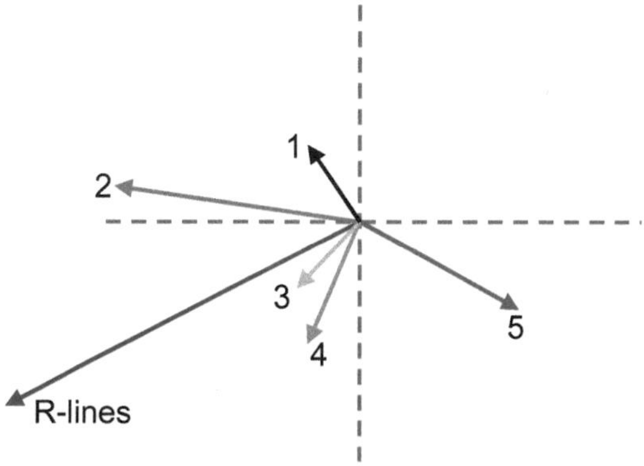

Fig. 3.6 Negative sequence global emission vectors of the individual lines and their resultant.

lines 2, 3 and 4 can be considered for transposition, where line 2 becomes the first candidate and then 4 and 3, respectively. Further, transposition of line 1 is not seen to lead to any improvement. This process has been verified by carrying out unbalanced load flow studies on the subtransmission network examined [14].

As for the case of asymmetrical lines, the network wide impact of the different unbalanced loads connected at various bus bars also can be represented by unique global emission vectors. The resultant of these emission vectors together with the individual global emission vectors can be used to determine the dominant contributors to the network wide voltage unbalance and assist in prioritising the load balancing process.

3.5.5 Voltage unbalance propagation and attenuation in power systems—The role of three-phase induction motors

Similar to other power quality disturbances, such as harmonics and flicker, voltage unbalance also can propagate from one location to another in power systems with or without much attenuation. For example, the voltage unbalance caused by electric traction systems that are usually supplied by dedicated medium voltage networks will propagate to the high voltage network via the connecting transformers. The same high voltage network connected to other load centres may experience the propagated voltage unbalance that is caused by the traction system. This propagation aspect was intrinsic in the meshed network analysis covered in Section 3.5.4. Much of the complex and detailed work in this regard (and in general on the management of voltage unbalance) can be found in Refs. [14,15,17].

With regard to voltage unbalance, the types of loads connected play an important role. Any installation that demonstrates a relatively smaller negative sequence impedance compared to its positive sequence impedance will act as a sink of negative sequence current whilst in general will tend to help attenuate pre-existing voltage unbalance. A prime example in this regard are the mains connected induction motors as covered in Section 3.4. In practical networks, noting that there is a vast collection of mains connected three-phase induction machines at LV and MV levels, their behaviour in the presence of voltage unbalance will have an impact on a network wide sense although such motors may experience negative consequences resulting from their terminal voltage unbalance. This voltage unbalance attenuation behaviour is demonstrated through a simulation where the 12.47-kV, 3.2187-km line of example of Section 3.5.3 is considered, however, with the line ideally transposed. The sending end phase voltages of the line are such that it is characterised by a negative sequence voltage unbalance of approximately 2%. The EMT [6] simulation outcomes are illustrated in Fig. 3.7 where the following are applicable:

- A constant impedance balanced three-phase load with $P = 4.2$ MW and $Q = 2.64$ MVAr is kept connected at the POE of the line.
- At 1.0 s, a group of eight (8) 4-kV, 200-kW three-phase induction motors is connected to the POE through a step down transformer having a negligible impedance.
- The motors are allowed to reach their steady operating behaviour and at 4 s, they are disconnected.
- At 5 s, a constant impedance balanced three-phase load with a P and Q equivalent to those of the group of motors is connected to the POE through the same transformer which supplied the motors.

In the time interval 0–1 s, during which the negative sequence voltage unbalance that propagates along the symmetrical line does not show any significant attenuation by comparing VUF_send with VUF_POE (ignoring the very short-term transient associated with the switching). Once the motors are switched on at 1 s, the period of acceleration of the motors to near full speed (W) is about 1 s over which the electromagnetic (T_e) torque of the motors demonstrates the usual oscillatory behaviour together with mechanical torque (T_m) which was set to demonstrate quadratic characteristic. VUF_POE is seen to decrease from 2% to about 1.8% during this period and stays at that value until the motors are disconnected at 4 s at which VUF_POE recovers back to 2%. It is clear that connection of the constant impedance load that is equivalent to the motor load at 5 s does not

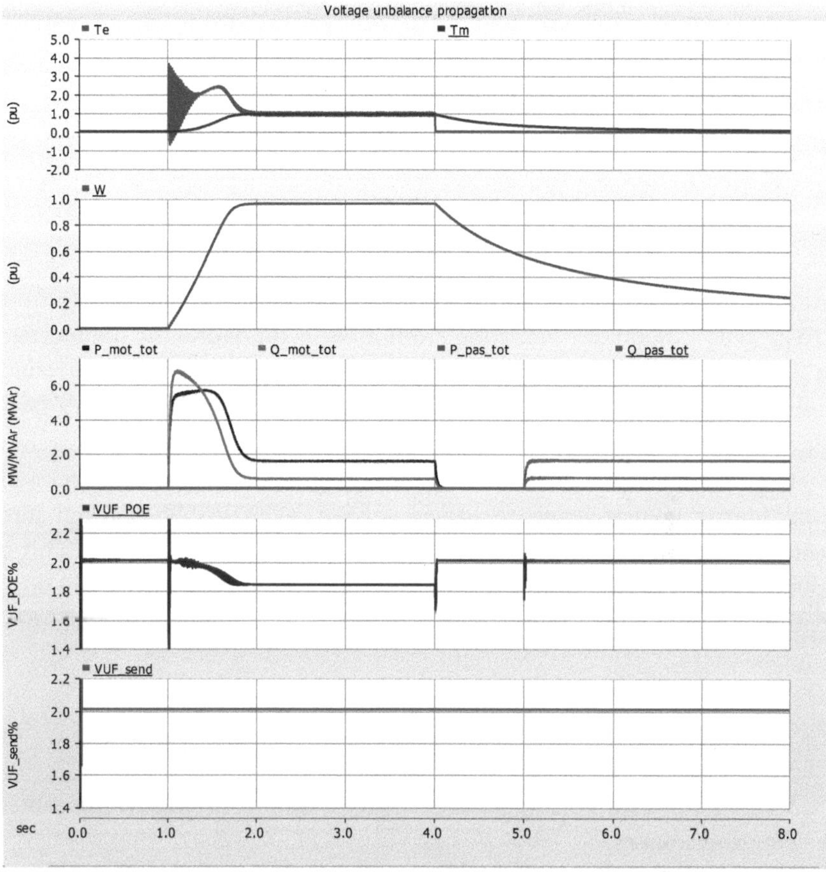

Fig. 3.7 Voltage unbalance propagation and attenuation—Simulation results.

demonstrate any contrasting change to the VUF_POE. In summary, these simulation outcomes clearly demonstrate the voltage unbalance attenuation behaviour of the three-phase induction motors in power systems.

3.5.6 Voltage unbalance network measurements and network planning levels

It is vital that voltage unbalance is measured paying attention to compliant instruments [5] employing transducers with acceptable accuracy levels across the three phases as the voltage unbalance levels refer to small percentages. Further, measurement durations and cumulative probability levels need to follow the recommended guidelines [4].

Table 3.2 Indicative planning levels [4].

Voltage level	Planning level %
MV	1.8
HV	1.4
EHV	0.8

Whilst voltage unbalance emission compliance with regard to an installation is carried out employing specific guidelines, it is interesting to note survey type outcomes originating from around the world, as summarised in Ref. [18] in relation to EHV, HV, MV and LV sites. Based on these results and with a view to coordinate with the compatibility level of 2% at LV level [9], indicative network planning levels given in Ref. [4] have been reached as presented in Table 3.2. In reaching these levels, voltage unbalance transfer from higher voltage levels to lower voltage levels has been taken into account in addition to the use of 1.4 as the summation exponent. This is covered in Refs. [4,18] and given in Eq. (3.27) which takes care of the summation of voltage unbalance caused by multiple voltage unbalance emitting sources.

$$VUF = \sqrt[\alpha]{\Sigma VUF_i^{\alpha}} \tag{3.27}$$

where

VUF = resultant voltage unbalance factor due to all voltage unbalance emitting sources.

VUF_i= voltage unbalance factor associated with the ith voltage unbalance emitting source.

α = summation exponent.

References

[1] A. Baggini, Handbook of Power Quality, John Wiley & Sons, 2008.
[2] A. von Jouanne, B. Banerjee, Assessment of voltage unbalance, IEEE Trans. Power Deliv. 16 (4) (2001) 782–790, https://doi.org/10.1109/61.956770.
[3] Gost 13109-97 Russian Standard, Power Quality Limits in Public Electrical Systems.
[4] IEC TR 61000-3-13:2008 Electromagnetic Compatibility (EMC)—Part 3-13: Limits—Assessment of Emission Limits for the Connection of unbalanced installations to MV, HV and EHV power systems.
[5] IEC 61000-4-30:2015+AMD1:2021 CSV Consolidated Version Electromagnetic Compatibility (EMC)—Part 4-30: Testing and Measurement Techniques—Power Quality Measurement Methods.
[6] PSCAD™ Version 5, Manitoba Hydro International.

[7] IEC 60034-26 Rotating Electrical Machines—Part 26: Effects of Unbalanced Voltages on the Performance of Three-Phase Cage Induction Motors.

[8] APQRC Tech Note 6, Voltage Unbalance. http://eisweb.adeis.uow.edu.au/apqrc/content/technotes/APQRC%20TN006-0210%20Voltage%20Unbalance.pdf.

[9] IEC 61000-2-2:2002 Electromagnetic Compatibility (EMC)—Part 2-2: Environment—Compatibility Levels for Low-Frequency Conducted Disturbances and Signalling in Public Low-Voltage Power Supply Systems.

[10] IEC 61000-2-12:2003 Electromagnetic Compatibility (EMC)—Part 2-12: Environment—Compatibility Levels for Low-Frequency Conducted Disturbances and Signalling in Public Medium-Voltage Power Supply Systems.

[11] IEC TR 61000-3-14:2011 Electromagnetic Compatibility (EMC)—Part 3-14: Assessment of Emission Limits for Harmonics, interharmonics, Voltage Fluctuations and Unbalance for the Connection of Disturbing Installations to LV Power Systems.

[12] J. Meyer, F. Moller, S. Perera, S. Elphick, General definition of unbalanced power to calculate and assess unbalance of customer installations, in: 2019 Electric Power Quality and Supply Reliability Conference and 2019 Symposium on Electrical Engineering and Mechatronics, PQ and SEEM 2019, 2019, pp. 1–6.

[13] CIGRE TB 468 Review of Disturbance Emission Assessment Techniques, 2011.

[14] P. Paranavithana, Contributions Towards the Development of the Technical Report IEC/TR 61000-3-13 on Voltage Unbalance Emission Allocation (PhD thesis), University of Wollongong, School of Electrical, Computer and Telecommunications Engineering, 2009. http://ro.uow.edu.au/theses/834.

[15] U. Jayatunga, Further Contributions to the Development of the Technical Report IEC/TR 61000-3-13:2008 on Voltage Unbalance Management Methodologies (Doctor of Philosophy thesis), School of Electrical, Computer and Telecommunications Engineering, University of Wollongong, 2013. https://ro.uow.edu.au/theses/3998.

[16] R.C. Dugan, Induction Machine Modeling for Distribution System Analysis—Test Case Description, in: 2005/2006 IEEE/PES Transmission and Distribution Conference and Exhibition, 2006, pp. 578–582, https://doi.org/10.1109/TDC.2006.1668560.

[17] T.D. Kahingala, Contributions to Analytical Methodologies for the Management of Voltage Unbalance in MV/HV/EHV Power Systems, Doctor of Philosophy thesis, School of Electrical, Computer and Telecommunications Engineering, University of Wollongong, 2020. https://ro.uow.edu.au/theses1/918.

[18] R. Koch, G. Beaulieu, L. Berthet, M. Halpin, International survey of unbalance levels in LV, MV, HV and EHV power systems: CIGRE/CIRED JWG C4.103 Results, in: 19th CIRED Conference, Paper 0892, May 2007.

CHAPTER 4

Impact and management of power system harmonics

4.1 Introduction

This chapter describes power system harmonic distortion. Whilst harmonic distortion is generally well understood, with academic papers on the topic being published early last century, continued evolution of technology with changing harmonic emission levels and equipment susceptibility has seen harmonic distortion remain a topic of interest for power quality engineers. The recent rapid increase in renewable energy generation, which is now almost exclusively interfaced with the supply network by means of power electronic converters (sometimes known as inverter-based resources—IBRs), has presented another challenge with respect to management of harmonic distortion in electricity supply networks.

In this chapter, sources of harmonic distortion are identified with the focus being on modern equipment as well as effects including impact on equipment behaviour. Whilst there are a few types of waveform distortion in power systems, this chapter predominantly focuses on integer harmonic distortion. Techniques for measurement and analysis of harmonic distortion are presented including the impact that transducers can have on measurements.

The chapter also provides an in-depth explanation of the at times complex IEC and IEEE guidelines which are used for management of harmonics, including application of the standards to large renewable energy generators (currently a topic of much conjecture). The limitations associated with the existing approaches used to assess harmonic compliance in electricity networks are also examined. The chapter concludes with a brief summary of strategies that can be used to mitigate harmonic distortion.

4.2 Definition of waveform (harmonic) distortion

Waveform distortion can be described as a variation in the shape of the waveform of the supply voltage away from the ideal sinusoidal waveform

Applied Power Quality
https://doi.org/10.1016/B978-0-323-85467-2.00008-1

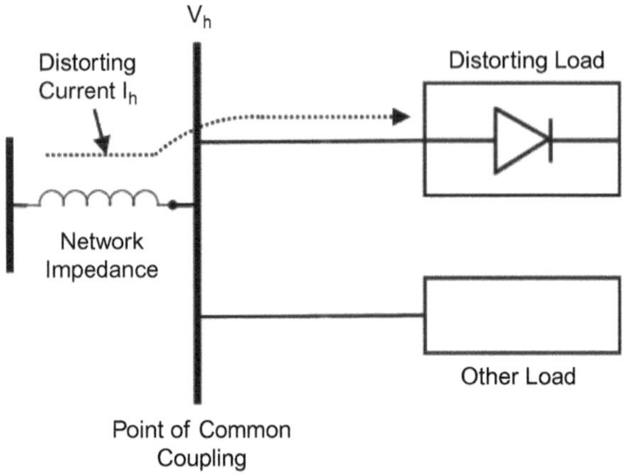

Fig. 4.1 Simple schematic showing source of waveform distortion.

and can be present in both voltage and current waveforms. This chapter considers waveform distortion that is caused by the interaction of nonsinusoidal currents with the supply network. Fig. 4.1 shows a simple schematic which demonstrates how distorted current (I_h) interacts with the network impedance to produce distorted voltage (V_h) at the point of common coupling of other loads within a power system. As a general principle, the magnitude of any waveform distortion is highest closest to the source of the distortion.

In modern power systems a large amount of the equipment connected to electricity distribution networks will draw nonsinusoidal current. This equipment can be referred to as distorting load. Possibly the best examples of such equipment are power conversion devices such as rectifiers and inverters. Another source of waveform distortion is equipment which will generally draw sinusoidal current but may draw distorted current under some operating scenarios. An example of this is the nonsinusoidal current drawn by saturated magnetic circuits.

4.2.1 Analysis of waveform distortion

Analysis of distorted waveforms is undertaken by resolving the waveform into spectral components. In its simplest sense, the Fourier series theorem states that any periodic waveform can be described by a sum of sinusoidal waveforms. Expressed another way, nonsinusoidal waveforms can be decomposed into a sum of the fundamental frequency waveform along with other sinusoidal waveforms whose characteristics, frequency, magnitude and phase angle, correspond to the distortion which is present. Whilst there are many forms of waveform distortion, the three main types observed in electricity supply networks are integer harmonic distortion, interharmonic distortion and high-frequency distortion. Each of these is examined as follows.

4.2.1.1 Integer harmonic distortion

Where the frequencies of the waveforms which make up the Fourier series of a distorted waveform are integer multiples of the fundamental frequency, these waveform components are termed harmonics. The harmonic order of any given waveform is simply the waveform frequency divided by the fundamental frequency. For example, in a 50-Hz power system, the 5th harmonic has a frequency of 250 Hz.

Integer harmonic waveform distortion is by far the most prevalent mechanism by which waveform distortion occurs in electrical power systems. Fig. 4.2 shows an example of a waveform which contains a 50-Hz fundamental component along with approximately 10% 5th harmonic distortion. The mathematical expression which describes the sum of the two sinusoidal waveforms (Fourier series) for the waveform shown in Fig. 4.2 is as follows:

$$\text{Amplitude} = 325\sin(2\pi ft) + 32.5(2\pi ft \times 5)$$

where

- f is the fundamental frequency (50 Hz in this example)
- t is time in seconds

In power system analysis, Fourier series is applied to distorted waveforms in order to determine the spectral components (harmonic frequency, magnitude and phase angle) of the waveform. Application of Fourier series to the waveform shown in Fig. 4.2 results in the harmonic spectrum shown in Fig. 4.3.

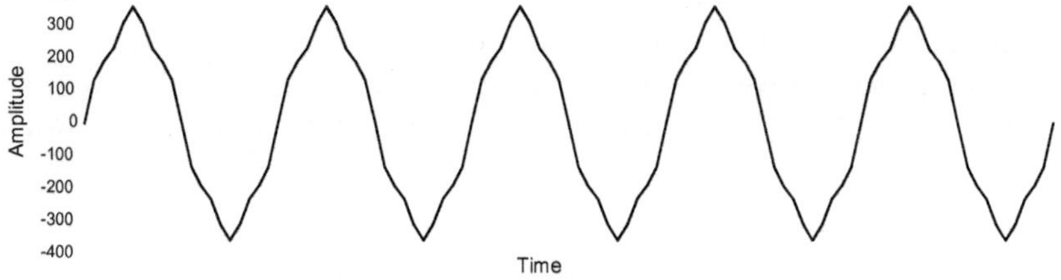

Fig. 4.2 Waveform with integer harmonic distortion.

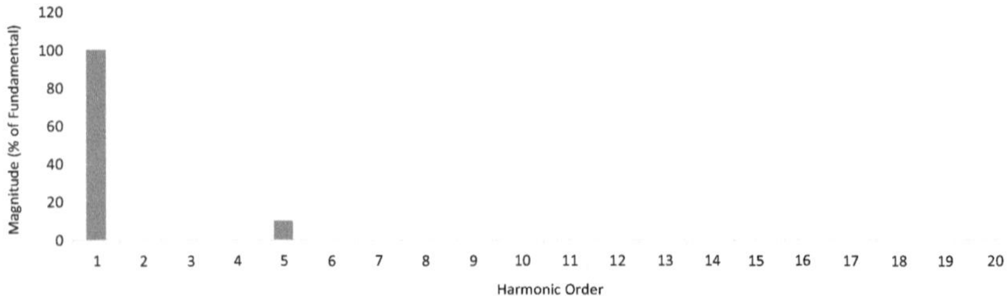

Fig. 4.3 Harmonic spectrum for waveform shown in Fig. 4.2

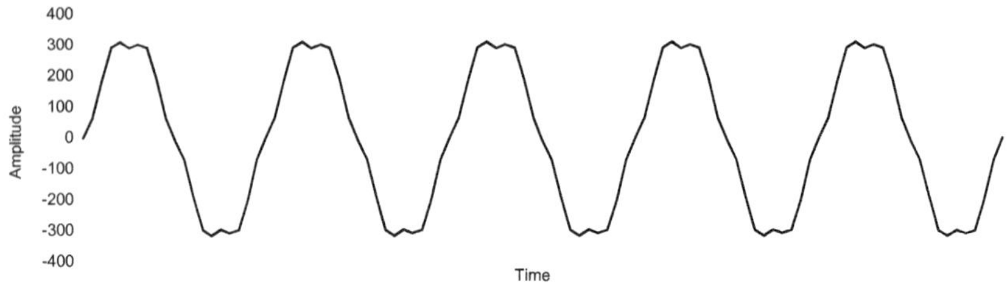

Fig. 4.4 Waveform with integer harmonic distortion (5th harmonic phase angle = 180 degrees).

The impact of harmonic distortion on the shape of the waveform is dependent on the phase angle of the spectral components. For example, if the phase angle of the 5th harmonic component present in the waveform in Fig. 4.2 is varied from 0 degrees (as shown) to 180 degrees the waveform will change shape from that shown in Fig. 4.2 to that shown in Fig. 4.4. The phase angle of the harmonic components may become important when considering the impact of the waveform distortion. In some cases, the harmonic components can cause the peak voltage to increase (as is the case in Fig. 4.2) which can increase the stress on components whilst flattening of the waveform (as is the case in Fig. 4.4) can change the conduction and associated current waveforms of power converters such as rectifiers.

4.2.1.2 Classification and harmonic phase sequence

Integer harmonics can be classified into three categories as follows:

- Even order harmonics—these are orders which are multiples of 2, i.e. 2nd, 4th, 6th, 8th, etc. Even order harmonic distortion is characterised by half-wave asymmetry.
- Odd order harmonics—these are orders corresponding to odd numbers but excluding multiples of 3 which are a special case, i.e. 5th, 7th, 11th, 13th, etc.
- Triplen harmonics—these are odd order multiples of three, i.e. 3rd, 9th, 15th, etc.

In the same way that the fundamental voltage waveforms can be decomposed into sequence components, the sine waves associated with harmonics can also be decomposed into sequence components and each harmonic order will have a 'characteristic' phase sequence. This phase sequence has important implications in harmonic management. The characteristic phase sequences for the harmonic classifications described before are as follows:

- Harmonics of order 1, 4, 7, 10 .... are positive sequence.
- Harmonics of order 2, 5, 8, 11 ... are negative sequence.
- Harmonics of order 3, 6, 9, 12 ... are zero sequence.

4.2.1.3 Interharmonic distortion

Interharmonic distortion is characterised by waveforms containing frequency components that are not integer multiples of the fundamental frequency. Waveforms with interharmonic distortion will not display half-wave symmetry, i.e. each half-cycle is different to the preceding half-cycle. Interharmonic distortion originates from devices such as static frequency converters, cycloconverters and arc furnaces. Fig. 4.5 shows an example of a waveform containing interharmonic distortion (5.5th harmonic or 275 Hz in this case).

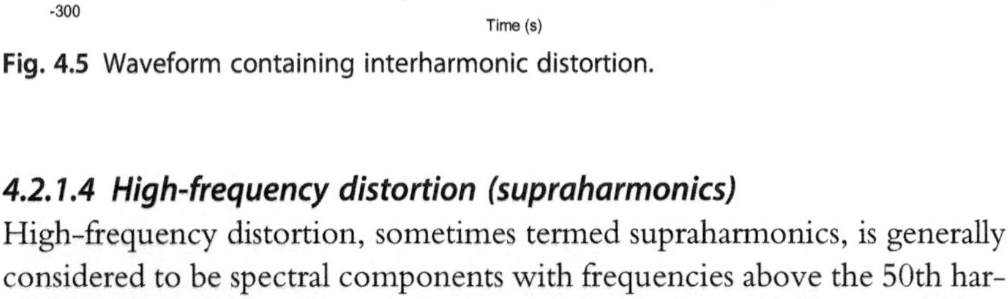

Fig. 4.5 Waveform containing interharmonic distortion.

4.2.1.4 High-frequency distortion (supraharmonics)

High-frequency distortion, sometimes termed supraharmonics, is generally considered to be spectral components with frequencies above the 50th harmonic (2.5 kHz for a 50-Hz system and 3 kHz for a 60-Hz system) and below 100 kHz. High-frequency distortion is caused by the switching of power conversion devices. As a general principle, the frequency of the switching components will reduce as the rating of the power converter increases, for example, a 10-kW inverter may have a switching frequency approaching 20 kHz, whilst a 2.5-MW inverter may have a switching frequency closer to 3 kHz. This variation in switching frequency is related to the characteristics and capability of the switching devices applicable at each voltage and power rating. Fig. 4.6 shows the current waveform for a 10-kW solar PV inverter measured using an oscilloscope. The high-frequency distortion on the waveform can be clearly observed. Fig. 4.7 shows the frequency spectrum for the waveform shown in Fig. 4.6. A frequency component close to 20 kHz is readily observable (Fig. 4.7).

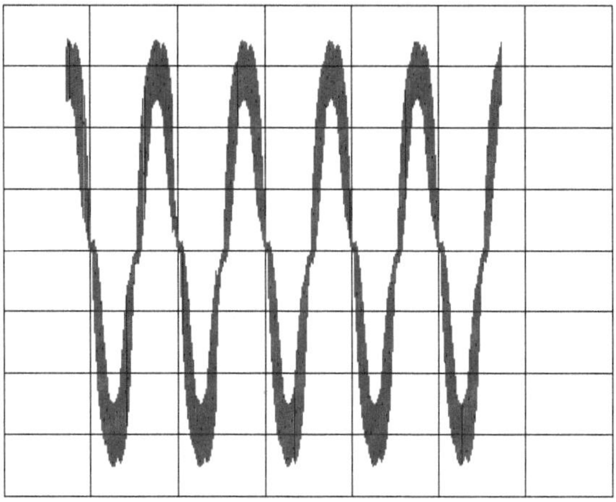

Fig. 4.6 Current waveform for a 10-kW soar PV inverter.

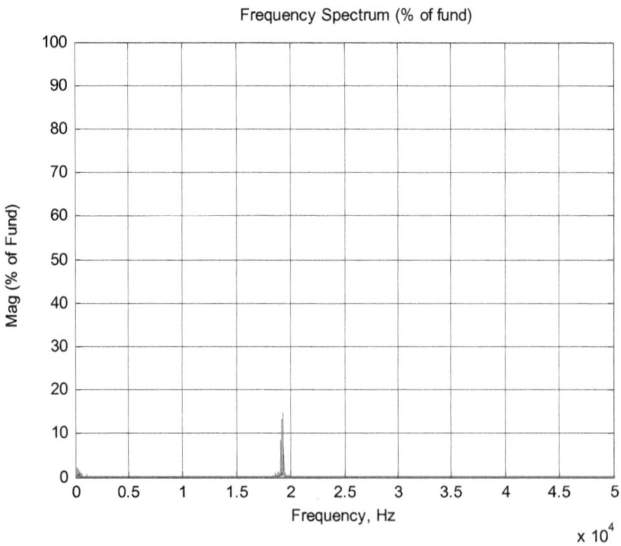

Fig. 4.7 Frequency spectrum of waveform shown in Fig. 4.6.

Whilst not a new phenomenon, high-frequency distortion is now receiving significant attention due to the prevalence of the power converters (inverters) associated with renewable energy generation. The impact of high-frequency harmonics has mostly been on communications systems, including power line carrier systems.

Whilst it is noted that waveform distortion can take many forms, by far the most prevalent and in some case problematic type of waveform distortion is integer harmonic distortion. For this reason, integer harmonic distortion is the focus of the remainder of the chapter.

4.3 Measurement and analysis of harmonic distortion

Measurement of harmonic and interharmonic distortion is achieved through the use of instrument (power quality monitors) capable of either measuring waveforms for postprocessing or performing an onboard Fast Fourier Transform (FFT) in order to resolve harmonic components. Chapter 8 provides an extensive account of the methods used for measurement of harmonics. As such, this section only provides high-level summary of the processes used to measure and analyse harmonic distortion.

4.3.1 Measurement standards

As detailed in Chapter 8, the internationally accepted standard which details the requirements of instruments used to measure power quality is IEC 61000-4-30. In the case of harmonics, IEC 61000-4-30 refers to IEC 61000-4-7 (Testing and measurement techniques—General guide on harmonics and interharmonics measurements and instrumentation, for power supply systems and equipment connected thereto) to specify the instrumentation requirements for measurement of harmonic distortion. The scope of IEC 61000-4-7 is as follows:

> *This part of IEC 61000 is applicable to instrumentation intended for measuring spectral components in the frequency range up to 9 kHz which are superimposed on the fundamental of the power supply systems at 50 Hz and 60 Hz. For practical considerations, this standard distinguishes between harmonics, interharmonics and other components above the harmonic frequency range, up to 9 kHz.*
> *This standard defines the measurement instrumentation intended for testing individual items of equipment in accordance with emission limits given in certain standards (for example, harmonic current limits as given in IEC 61000-3-2) as well as for the measurement of harmonic currents and voltages in actual supply systems. Instrumentation for measurements above the harmonic frequency range, up to 9 kHz is tentatively defined (see Annex B) [1].*

Whilst it is beyond the scope of this text to delve into the details of IEC 61000-4-7, the standard specifies requirements such as accuracy and classes of instrument and windowing requirements for FFT calculation. The standard effectively ensures that if two instruments which are compliant with the standard measure the same signal, the outputs will be equal within the limitations of the accuracy class.

4.3.2 Transducers

Understanding of the capabilities and characteristics of the transducers which must be used to reduce large voltages and currents to values which can be measured by power quality monitors is an important part of measurement of harmonics. Once again, Chapter 8 provides significant details on transducer performance. For the purposes of this section, the following high-level summary is sufficient:

- For voltage transducers or voltage transformers (VTs), the frequency response, i.e. the highest harmonic order that can be effectively measured using the device, varies with type and voltage level. Capacitive voltage transformers (CVTs), which are used at high voltage, are not suitable for measurement of harmonics due to the fact that they are a low-pass filter by design. Compensator devices can be retrofitted to CVTs to enable harmonics measurement capability. The suitability of magnetic VTs for harmonics measurement can vary widely based on factors such as voltage level, type and burden. As a general principle, the highest harmonic order that can be measured accurately decreases as nominal voltage increases. Specialised voltage divers, whether they be resistive, capacitive or some combination of the two, generally have sufficient bandwidth to provide accurate measurement of all the harmonic orders on interest.
- Measurement type current transducers or transformers generally have sufficient frequency response to measure harmonics to the 40th or 50th order.

4.3.3 Analysis techniques

An instrument (or measurement system) capable of measuring harmonics and/or interharmonics will generally produce values for each harmonic order. These values can include magnitudes and phase angles for voltage, current and power. The values for magnitude can be expressed as absolute magnitudes (e.g. voltage parameters given in volts) or may be expressed as percentages of a user-defined value such as the nominal or fundamental value. Particular care should be taken when examining magnitudes expressed as a percentage of the fundamental value as harmonic values may not decrease linearly with the fundamental value. As such, magnitudes expressed as a percentage of the fundamental value may appear large whilst the absolute magnitude may be relatively small. This is often the case for current THD.

In some cases it is useful to combine all measured harmonic orders into a single value (or index) in order to achieve a simple understanding of the harmonic performance. There are two well-known indices that are used for this purpose: Total Harmonic Distortion (THD) and Total Demand Distortion (TDD).

THD is expressed as the sum of the squares of the harmonic components expressed as a percentage of the fundamental, i.e.

$$\mathrm{THD} = \frac{\sum_{2}^{n} \mathrm{Value}_n}{\mathrm{Fundamental}}$$

where

- n is the harmonic order
- Value_n is the measured value for the nth harmonic order
- Fundamental is the fundamental value

THD can be calculated for both voltage and current. TDD only applies to current and replaces the time-varying fundamental value in the THD calculation with a fixed demand value, i.e.

$$\mathrm{TDD} = \frac{\sum_{2}^{n} I_n}{\text{Maximum demand}}$$

where

- n is the harmonic order
- I_n is the measured value of current for the nth harmonic order
- Maximum demand is a user-defined value which generally corresponds to the maximum demand load current of the installation or equipment being assessed.

The most recent version of IEEE standard 2800 (IEEE Standard for Interconnection and Interoperability of Inverter-Based Resources (IBRs)

Interconnecting with Associated Transmission Electric Power Systems (2022)) introduces another metric for combined harmonic distortion—total rated current distortion (TRD). TRD is defined as follows:

$$\text{TRD}\ (\%) = \frac{\sqrt{I_{\text{rms}}^2 - I_1^2}}{I_{\text{rated}}} \times 100\%$$

where

- I_1 is the fundamental frequency current as measured at the reference point of applicability (RPA).
- I_{rated} is the *IBR plant* rated current capacity based on *IBR plant* MVA rating at the RPA (as stated in the interconnection request; may be equal to continuous rating of the IBR plant).
- I_{rms} is the root mean square of the *IBR plant* current, inclusive of all frequency components up to 50th order, as measured at the RPA; measurement of harmonics to the 50th order requires metres compliant to IEC 61000-4-30 Class A.

4.4 Sources of harmonic distortion

Distortion of the voltage waveform is caused by the interaction between distorted current waveforms and network impedance according to Ohm's law. In simple terms, the following applies:

$$V_n = I_n \times Z_n$$

- V_n is the harmonic voltage for the nth order harmonic.
- I_n is the harmonic current for the nth order harmonic.
- Z_n is the network impedance for the nth order harmonic.
- n is the harmonic order (2,3,4,….).

Sources of distorted current can include loads as well as electricity network components. As discussed before, loads which draw distorted current waveforms are often termed distorting loads or nonlinear loads. It is important to note here that from the perspective of waveform distortion, no distinction needs to be made between the distorted current drawn by either loads or generators, i.e. both can be correctly referred to as distorting loads.

Specific examples of sources of distorted current waveforms include the following:

- Power electronic loads designed for power conversion such as rectifiers, inverters and variable speed drives. These devices can be a source of both low order (i.e. up to the 50th harmonic order) and high-frequency harmonics (distortion at switching frequencies).

- Magnetising current for devices such as motors and transformers can be highly distorted; however, the magnitudes of these currents are generally small compared to the rating of the device. However, if a magnetic circuit is subject to saturation, the magnitude of the distorted current can increase by at least an order of magnitude.
- Arcing devices such as arc furnaces and discharge lighting systems.

Chapter 7 provides a detailed analysis of the characteristics of, and current waveforms drawn by, a range of the most common equipment connected to the power system.

Another source of waveform distortion is communication signals which are superimposed onto the power system by network operators for applications such as load control. An example of this is audio frequency injection control (AFIC) signalling which is used in a number of locations worldwide to control (turn on and off) loads such as hot water heating and street lighting. These signalling systems are designed for demand management and can allow consumers to access lower priced electricity tariffs at specific times of the day. The frequency range for AFIC is generally between 200 Hz and 1100 Hz. AFIC signalling systems have been observed to be responsible for a number of undesirable impacts on network equipment particularly if magnitudes are amplified by resonance (the impact of resonance on the magnitude of harmonic distortion is discussed in detail in Section 4.3.1). These impacts include audible noise through speakers as well as noise induced by high-frequency vibration of the laminations in motors such as those found in ceiling fans. One of the more unusual impacts of high levels of relatively high-frequency waveform distortion (e.g. the 1050-Hz ripple injection signalling frequencies used in a number of countries) is maloperation of equipment which relies on zero crossings for correct operation. Examples of this equipment include digital clocks which count zero crossings to keep time. If the AFIC component of the waveform distortion is high enough, the waveform may exhibit multiple zero crossings for each fundamental frequency cycle. An example of this is shown in Fig. 4.8 which shows one cycle of the fundamental 50-Hz waveform which includes 1050-Hz distortion with a magnitude of 10% of the fundamental and a phase angle of 180 degrees. Multiple zero crossings can clearly be observed in Fig. 4.8. In the case of devices which count zero crossings to keep time, when exposed to this waveform they will count additional zero crossings and as such will 'gain' time or run fast. Hence this phenomenon is often termed 'racing clocks'.

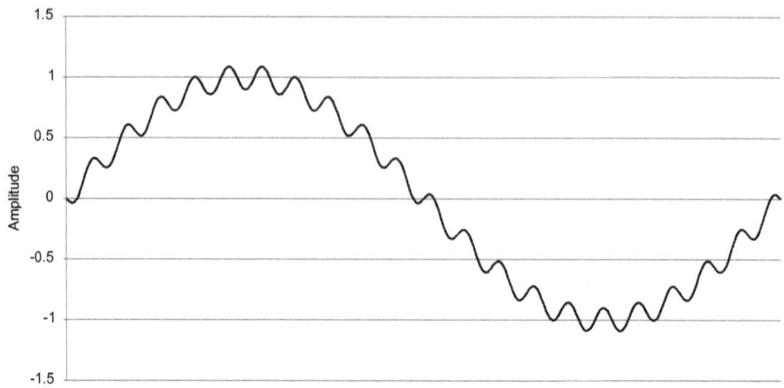

Fig. 4.8 Example of waveform displaying multiple zero crossings due to 1050-Hz distortion.

Other impacts of this type of waveform distortion include maloperation of power factor measurement devices which use the time difference between the zero crossings of the voltage and current waveforms to calculate power factor.

4.4.1 Resonance

A network resonance condition can result in what otherwise might be benign magnitudes of waveform distortion to become highly problematic. There are two categories of resonance that must be considered: series and parallel resonance. Of these two categories, parallel resonance is generally more important when considering harmonic distortion. Before further consideration of the impact of network resonance on harmonic distortion, it is necessary to consider how network impedance varies with frequency.

4.4.1.1 Frequency dependence of network impedance

Considering an inductor, its inductive reactance X_L (in Ω) is given by the equation:

$$X_L = 2\pi f L$$

where

- X_L is the inductive reactance.
- f is the frequency in Hz.
- L is the inductance in H.

Considering the above from a harmonics perspective, the value of the inductive reactance X_{hL} at any particular harmonic order h is given by

$$X_{hL} = hX_L$$

From the above it can be seen that inductive reactance increases as frequency increases.

Considering a capacitor, its capacitive reactance X_C (in Ω) is given by the equation:

$$X_C = \frac{1}{2\pi fC}$$

where

- X_C is the capacitive reactance
- f is the frequency Hz
- C is the capacitance in F

Considering the above from a harmonics perspective, the value of X_{hC} at any particular harmonic order h is given by:

$$X_{hC} = \frac{X_C}{h}$$

From the above it can be seen that capacitive reactance decreases with frequency.

The above has illustrated the influence of frequency on network impedance. The following are general observations concerning this phenomenon:

- At low frequencies, the impedance of the power system is dominated by the low inductive impedance of transformers and transmission lines.
- At high frequencies it is dominated by the low capacitive impedance of power factor correction capacitors.
- There is an intermediate range of frequencies where the capacitive and inductive effects can combine to give a very high impedance (as discussed further later). In this case, a small harmonic current within this frequency range can give rise to a very high and undesirable harmonic voltage. This is the condition which is called parallel resonance.

4.4.1.2 Series resonance

Series resonance occurs when the values of the inductive and capacitive impedance connected in series in part of a network are equal. Recalling the convention of inductive impedance being assigned a positive value and capacitive impedance a negative value, when the inductive and capacitive impedances are equal they will cancel resulting in a much lower impedance than would otherwise be the case. The implication here is that any voltage across a series resonant circuit will result in a much larger current than would be the case if the resonance were not present. This increased current can exceed device and component ratings and as such can lead to failure.

4.4.1.3 Parallel resonance

A parallel resonance will occur when the impedance seen by a source of harmonic current consists of an inductor and a capacitor connected in parallel. The shunt capacitors used for power factor correction banks are one of the major sources of this circuit configuration. As stated, power factor correction capacitors are connected in shunt and from the perspective of the load, or source of harmonic current, appear as a parallel circuit element to the upstream network impedance which is predominately inductive in distribution networks. This circuit configuration is illustrated in the circuit diagram shown in Fig. 4.9. This combination of circuit components can lead to a much higher than anticipated impedance at a frequency termed the resonant frequency. If current with a frequency close to the resonant frequency is present, a very high level of voltage distortion may be observed.

Considering the circuit shown in Fig. 4.9, the impedance, Z_h, as seen by the distorting load (sometimes referred to as the load point or driving point impedance), consists of the inductive impedance of the upstream network in parallel to the capacitive impedance of the power factor correction capacitor and is given by the equation:

$$Z_h = jhX_L \Big\| \frac{-jX_C}{h}$$

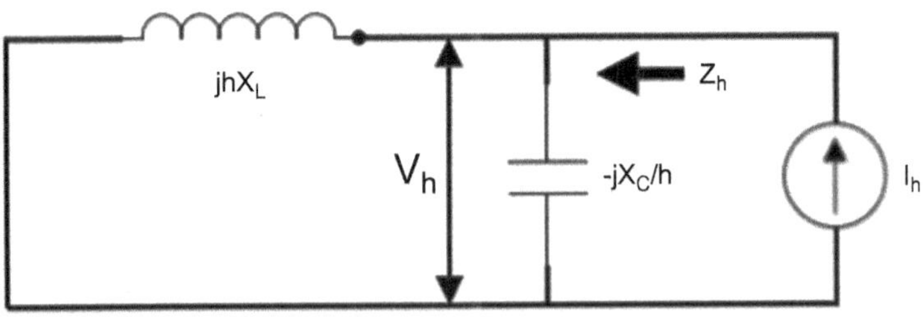

Fig. 4.9 Example of circuit that will exhibit parallel resonance.

This can be re-expressed as follows:

$$|Z_h| = \frac{hX_L X_C}{h^2 X_L - X_c}$$

The right-hand side of the previous equation will tend towards infinity, i.e. the value of Z_h will become very high and is limited only by whatever resistance is present in the circuit when the dominator equals zero, i.e.

$$h^2 X_L - X_c = 0$$

Rearranging the above gives the resonant value in terms of harmonic order h_{res} as follows:

$$h_{res} = \sqrt{\frac{X_C}{X_L}}$$

Recalling that the harmonic voltage for the hth harmonic order is given by

$$V_h = I_h \times Z_h$$

If there is harmonic current with frequency close to the resonant frequency, a much large harmonic voltage will occur than would otherwise be the case.

The size of a capacitor bank is usually given in terms of the reactive power generated, Q_C. Let the fault level at the point of capacitor connection be FL. If both are measured in consistent units, e.g. MVAr and MVA, then a resonance will occur at harmonic order (h_{res}) as follows:

$$h_{\text{res}} = \sqrt{\frac{FL}{X_c}}$$

Examples

In order to better understand the concepts of the dependence of impedance on harmonic distortion and the impacts of parallel resonance, a series of examples are presented as follows. The examples consider an 11-kV zone substation with a fault level of 220 MVA.

In the first instance, consider the situation without any power factor correction capacitors, as is the case for the circuit shown in Fig. 4.10 is

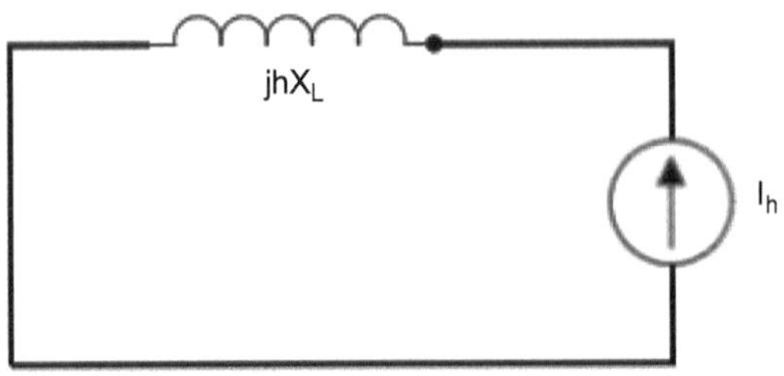

Fig. 4.10 Circuit with simple inductive upstream impedance.

considered. In this case the upstream impedance (X_L in Ω) at the fundamental frequency is given by:

$$X_L = \frac{\text{Voltage (kV)}^2}{\text{Fault Level (MVA)}} = \frac{11^2}{220} = 0.55\ \Omega$$

If it is assumed that the upstream impedance is effectively inductive, recalling that the relationship between inductive impedance and harmonic order is given by $X_{hL}=hX_L$ as shown in Fig. 4.11, the impedance of the circuit increases linearly with harmonic order.

If a power factor correction capacitor is connected to the circuit shown in Fig. 4.10, the result is the circuit shown in Fig. 4.9. For the purposes of this example, in the first instance, let the rating of the power factor correction capacitor be 2.5 MVAr. The capacitive impedance (X_C) of this power factor capacitor is as follows:

$$X_C = \frac{\text{Voltage (kV)}^2}{\text{Capacitance (MVAr)}} = \frac{11^2}{2.5} = 48.4\ \Omega$$

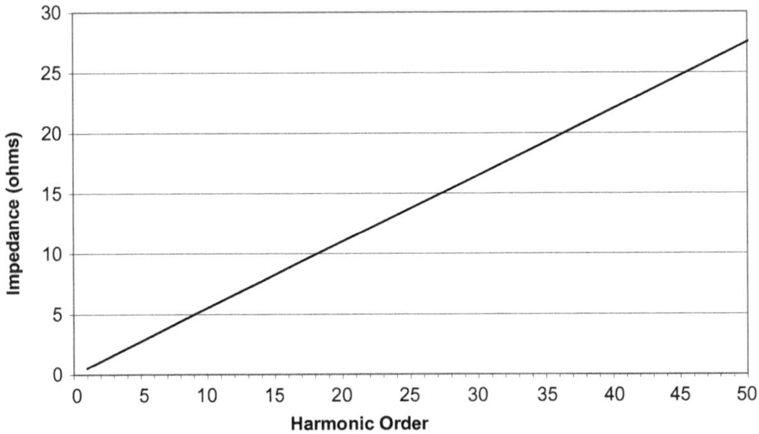

Fig. 4.11 Relationship between impedance and harmonic order for a simple inductive impedance.

The absolute value of the impedance of this parallel combination of upstream inductive impedance and power factor correction capacitor is given by:

$$|Z_h| = \frac{hX_L X_C}{h^2 X_L - X_c} = \frac{h \times 0.55 \times 48.4}{h^2 \times 0.55 - 48.4}$$

Values of Z_h for 50 harmonic orders are shown in Fig. 4.12 where it can be seen that there is a resonance between the 9th and 10th harmonic orders. Recalling the relationship between the harmonic order (h) at which resonance will occur is shown as follows, for this example, the harmonic order at which resonance occurs is 9.38. In this scenario, if there is harmonic current at or near the 9th or 10th harmonic orders, harmonic voltages considerably larger than those that would otherwise be expected will occur.

$$h = \sqrt{\frac{FL}{X_c}} = \sqrt{\frac{220}{2.5}} = 9.38$$

In many cases, power factor correction capacitors are deployed in steps in order to provide some control granularity. Fig. 4.13 shows the impact on impedance of four capacitor steps (2.5, 5, 7.5 and 10 MVAr). The harmonic order of the resonant frequency for each step is presented in Table 4.1. As shown, for this example, resonances occur at harmonic orders close to the 5th and 7th. It is well known that 5th and 7th are orders that are characteristics of bridge rectifier circuits and as such it can

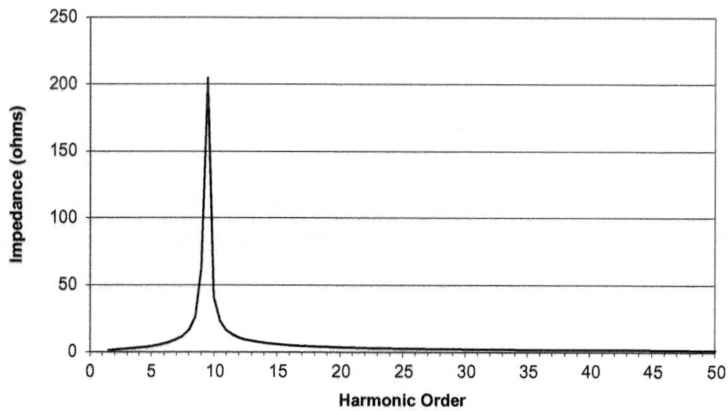

Fig. 4.12 Impedance for parallel combination of upstream inductive impedance and 2.5-MVAr power correction factor capacitor.

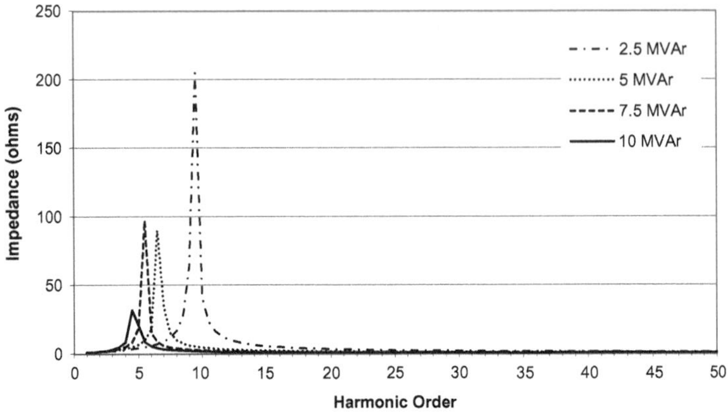

Fig. 4.13 Impedance for parallel combination of upstream inductive impedance and power factor capacitors.

Table 4.1 Harmonic order of resonant frequencies for capacitor steps.

Capacitor step (MVAr)	Harmonic order of resonant frequency (h_{res})
2.5	9.4
5	6.6
7.5	5.4
10	4.7

be reasonably expected that some value of 5th and 7th harmonic current will be present in the network. As such, the resonance created by some of these power factor correction steps may be problematic. The next section describes a method of avoiding resonance at problematic harmonic orders.

4.4.1.4 Mitigating the impacts of parallel resonance

Parallel combination of an inductance and a capacitance can form a resonant circuit. Whilst network and/or load impedance will reduce (damp) resonance it is impossible to eliminate resonance altogether. If it is accepted that resonance will occur, one method of reducing the impact of resonance on harmonic voltage magnitudes is to ensure that the resonant frequency does not correlate with a frequency at which significant magnitudes of harmonic current are likely to be observed, i.e. ensure that the resonant frequency does not align with the likely frequency of harmonic current emissions. This can be achieved by adding an inductor in series with the capacitor in order to 'tune' the circuit such that the resonant frequency occurs at a user-defined value. This inductor is called a detuning inductor. A schematic of the circuit including the detuning inductor represented by reactance (X_D) is shown in Fig. 4.14.

The series branch consisting of X_C and X_D in Fig. 4.14 has a series resonance when the following is true:

$$hjX_D = \frac{-jX_C}{h}$$

Rearranging the previous equation enables X_D to be calculated in terms of h as follows:

$$X_D = \frac{X_C}{h^2}$$

h is typically selected to be a value below the lowest harmonic order for which significant harmonic current is expected. For example, in a reasonably balanced medium voltage network with loads connected via star-delta transformers, the first harmonic order where significant harmonic current would be expected would be the 5th (as 3rd harmonic current is trapped by the transformer delta winding).

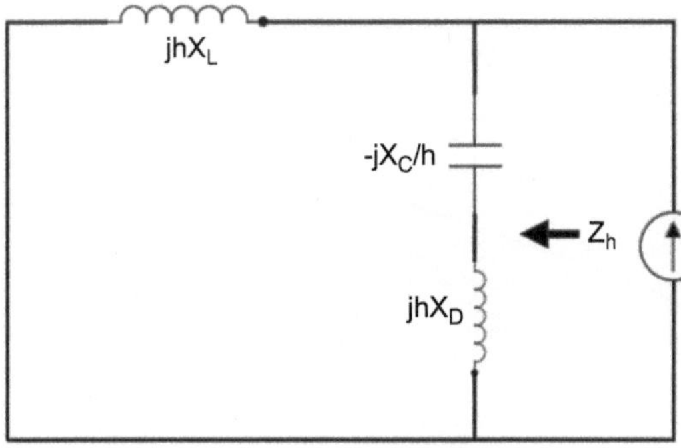

Fig. 4.14 Circuit containing detuning inductor (X_D).

Example

Using the same data as for the example presented in Section 4.4.1.3 this example will work through the process of selecting a detuning inductor in order to mitigate the impacts of the resonant circuit for the 5-MVAr capacitor. For the purposes of this example, the proposed resonant frequency after detuning is 3.5. This value is below the lowest harmonic order where significant harmonic current would be expected (5th harmonic). X_D is calculated as follows:

$$X_D = \frac{24.2}{3.5^2} = 1.98$$

Selecting this value of X_D ensures that the series branch consisting of X_C and X_D appears as an inductive impedance for harmonic orders greater than h. This is illustrated graphically in Fig. 4.15. In this figure, negative impedance indicates that the circuit is capacitive whilst positive impedance indicates inductive impedance. As stated, beyond h, the impedance of the branch is inductive and resembles the impedance shown in Fig. 4.11.

The total impedance of the circuit shown in Fig. 4.14 when h is selected to 3.5 (giving $X_D=1.98$) is shown in Fig. 4.16. Here it can be seen that the resonant frequency has been moved from close to the 5th harmonic to the 3.5th harmonic. It can also be seen that the circuit is inductive beyond the value of n. In addition, it can easily be shown that X_D will be significantly larger than X_L and as such, for harmonic orders above n, the circuit impedance can be reasonably approximated to be X_L.

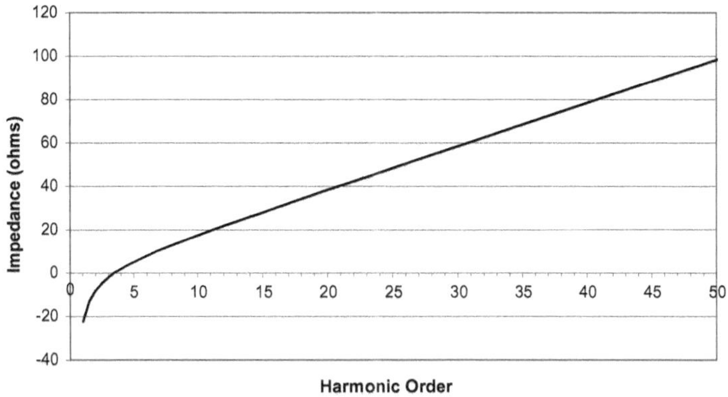

Fig. 4.15 Impedance of branch circuit consisting of X_C and X_D.

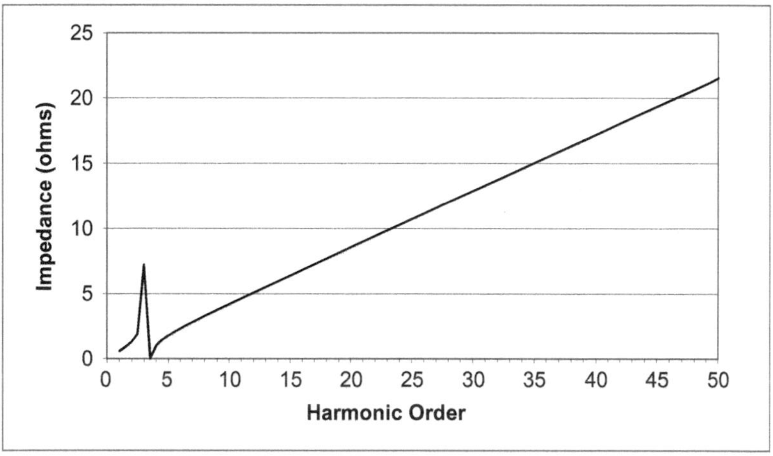

Fig. 4.16 Total impedance of the circuit shown in Fig. 4.14 for $n=3.5$.

Whilst detuning inductors can be used to mitigate the impacts of parallel resonance, it must be noted that the rating of the capacitor must be increased. This is due to the fact that the detuning inductor reduces the series impedance of the shunt circuit containing the capacitor and as such additional current will flow for any given voltage.

4.4.2 Characteristics of harmonic distortion in electricity distribution networks

The studies published in Refs. [2,3] provide insight into the magnitude and behaviour of harmonic voltages in electricity supply networks. Both of these studies utilise data collected from Australian electricity distribution networks.

The study presented in Ref. [2] shows that voltage THD levels are quite low and well within limits imposed by Australian standards, as demonstrated by Fig. 4.17 (the limit for voltage THD is 8%).

The study also indicated that voltage THD magnitudes were trending downward. The hypothesis for this downward trend is that equipment emissions are improving and that a significant volume of distorting load has been removed from the network due to many large manufacturing plants being relocated to overseas locations.

A comparison has been made between voltage THD levels observed on Australian networks with voltage THD levels measured on a number of other networks around the world. The results of this comparison are mixed,

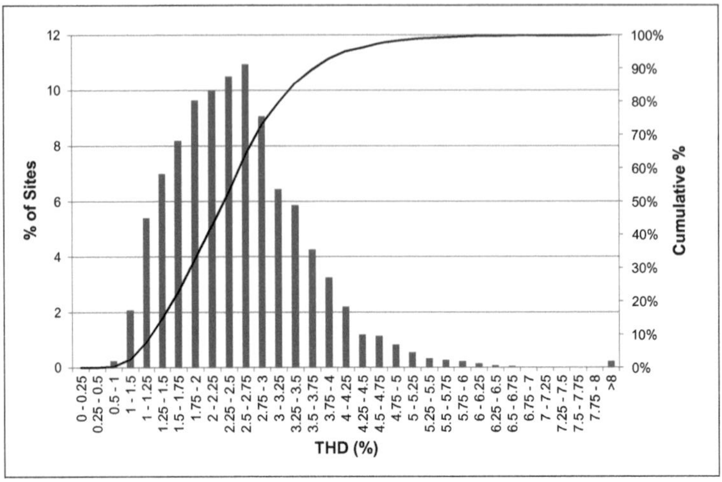

Fig. 4.17 Voltage THD magnitudes in low voltage networks from Ref. [2].

with Australian levels being lower than those observed in Europe but higher than those observed in other parts of the world such as Taiwan and India.

Data shown in Ref. [3] provides a comparison of voltage THD performance at strong and weak low voltage sites. For the purposes of this study, a strong site is one which is close to the distribution transformer whilst a weak site is a site remote from the distribution transformer. This comparison of voltage THD magnitudes is reproduced in Fig. 4.18 where it can be seen that voltage THD levels are higher at weak sites than strong sites. This result, that is, harmonic distortion magnitudes will increase as distances from the distribution transformer increase, needs to be considered when monitoring of power quality is undertaken (Fig. 4.18).

4.5 Impact of harmonic distortion

The impact of harmonics on equipment and electricity network components can highly vary and ranges from difficult to observe long-term thermal impacts such as loss of life in devices through to catastrophic failures. This section provides an overview of the impact of harmonics on common customer and network equipment.

One of the key impacts of the current drawn by distorting loads is an increase in the overall magnitude of the current (in amperes) flowing in networks and customer installations. In some cases, this can have thermal rating implications and can significantly increase losses. A special case of this

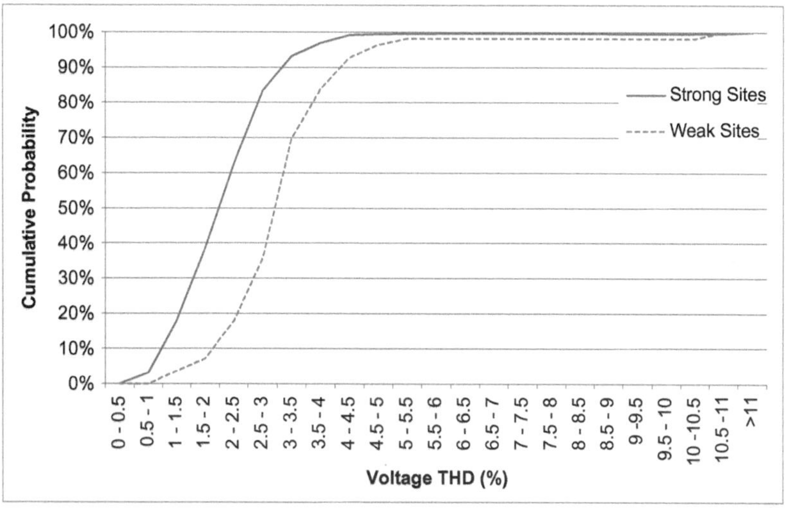

Fig. 4.18 Comparison of THD magnitudes at strong and weak low voltage sites.

phenomenon is that seen for triplen harmonics. The dominant sequence of these harmonic orders is zero sequence. As such, three-phase harmonic currents at these orders will add in neutral conductors as opposed to fundamental currents which will cancel in a balanced three-phase system. Before the widespread use of distorting loads, and before the impact of harmonics was well understood, the neutral conductor in three-phase installations was often much smaller in size than the phase conductors based on the assumption that the neutral current would be negligible due to cancellation. Once distorting loads became common there were many examples of overloading of neutral conductor due to triplen harmonics. This problem is now rare as most installations now have neutral conductors at least the same size as the phase conductors.

4.5.1 Impact on customer equipment

4.5.1.1 Electronic loads

The front end of most electronic loads consists of a bridge rectifier followed by a DC bus (filter capacitor). The high-frequency components of voltage harmonics can lead to loss of life in these capacitors due to their low impedance at high harmonic frequencies which leads to higher capacitive currents and higher heating losses.

Another impact of voltage distortion on electronic equipment is that in the presence of distorted supply voltages, electronic equipment will draw

Table 4.2 Components of waveforms of voltages used to evaluate impact of supply voltage on harmonic current emissions.

	Waveform 1		Waveform 2	
Harmonic order	Magnitude (% of fundamental)	Phase angle (degrees)	Magnitude (% of fundamental)	Phase angle (degrees)
3	1	0	2	0
5	3	180	4	180
7	1	50	1.5	50

currents containing harmonic components that would not otherwise be present if the device was supplied with a sinusoidal waveform. These harmonic components can be at unexpected orders. As an example, the impact of distortion of the supply voltage on the harmonic current emissions from equipment measurements of a compact fluorescent lamp (CFL) is presented. The lamp was supplied with a sinusoidal waveform as well as the waveforms presented in Table 4.2. The resulting current waveforms and current harmonic spectra are shown in Figs 4.19 and 4.20, respectively. It can clearly be seen that supplying the CFL with waveforms containing harmonic distortion impacts the shape of the waveform and hence the harmonic spectrum. In this case, the presence of low order harmonic distortion has significant impacts on the magnitude of low order harmonic current emission.

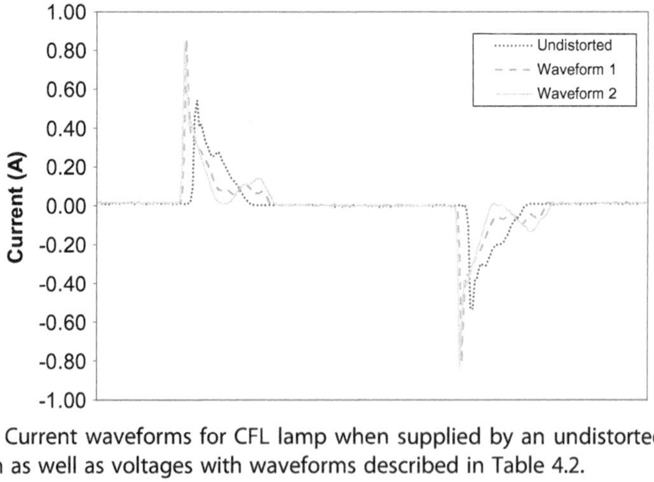

Fig. 4.19 Current waveforms for CFL lamp when supplied by an undistorted voltage waveform as well as voltages with waveforms described in Table 4.2.

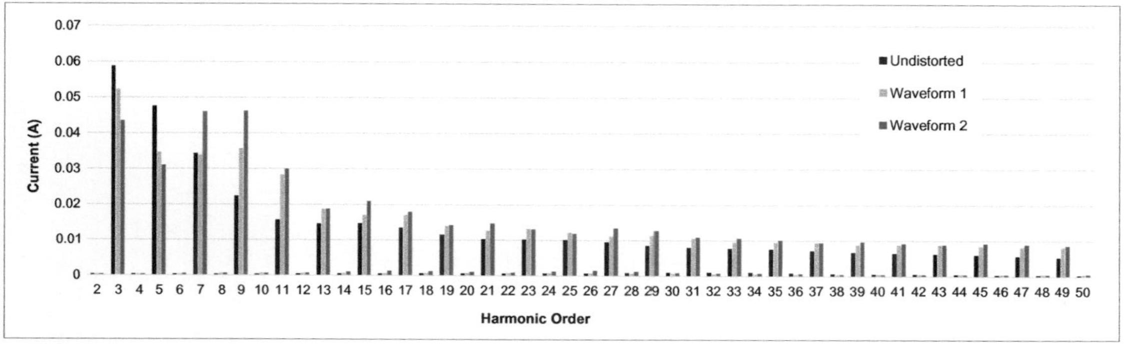

Fig. 4.20 Current harmonic spectra for CFL Lamp when supplied by undistorted voltage waveform as well as waveforms described in Table 4.2.

4.5.1.2 Motor loads

The main impact of voltage harmonics on motor loads has been identified as additional losses in single- and three-phase motors and loss of life in single-phase motors. In simple terms, motors present an inductive impedance to the network. When supplied with a voltage containing harmonic content, this impedance will draw an inductive harmonic current in addition to the fundamental current. Low-frequency harmonics are the most important in this process. These harmonic currents do not assist motor operation and manifest as a loss by inducing additional heating in the motor. In addition, harmonic orders which are predominately negative sequence create counter-rotating magnetic fields in three-phase motors which again increase heating (similar to the impact of negative sequence voltage unbalance). The additional heating associated with the harmonic current is generally small and tends to have little impact on three-phase motor insulation due to the fact that the motors are generally oversized for their load. However, this may not be the case for single-phase motors which are generally designed to closely match the load. As such, the additional heating caused by harmonic currents in single-phase motors is more likely to lead to degradation of insulation and associated loss of life.

4.5.1.3 Power factor correction capacitors

As described in Section 4.4.1.1, the impedance of capacitors decreases as harmonic increases. The implication of this is that power factor correction capacitors can present a very low impedance path for harmonic current. If the additional harmonic current exceeds the capacitor rating, the impacts can vary from loss of life through to catastrophic failure.

4.5.2 Economic impact of harmonics

4.5.2.1 Economic impact to consumers

Costs for voltage harmonics are generally divided into costs due to additional losses and costs due to equipment maloperation due to waveform distortion. In Ref. [4], the costs have been divided into direct and indirect costs as follows:

Direct costs

- Damaged components (e.g. capacitor banks)
- Damage to parts of the installation (e.g. cables, transformers)
- Costs due to loss of production
- Loss of staff productivity
- Costs of repairs

Indirect costs

- Additional loading of components
- Additional energy losses
- Ineffective use of the current-carrying capacity of the installation
- Reduced lifetime of components
- Less effective functioning of protection components

In addition to the above, [5,6] state that the costs associated with voltage harmonics for customers and networks are as follows:

- Iron losses (eddy currents)
- Costs associated with resonance phenomena
- Derating of network equipment, mainly of transformers
- Malfunction of equipment and industrial processes
- Noise and vibration
- Reduction in motor torque
- Decreased power factor
- Decreased performance of television sets and relays
- Inaccurate readings from induction watt-hour metres.

The costs associated with harmonic vary widely in the literature. Some studies give extremely high cost for harmonics whilst others suggest that harmonics are a relatively benign problem. In Ref. [7] it is also stated that for harmonics the focus is on the 5th, 7th and 11th orders. Energy loss due to harmonics is claimed to be about 50% greater.

Four case studies are cited in Ref. [6] which estimate the cost of harmonics to be as follows:

- In France, the cost of harmonic mitigation could be as high as € 25 billion. Given a GDP of € 1400 billion, this equates to 1.8% of GDP.
- In Quebec, heating losses and ageing due to harmonics have been estimated as $650 million. Given a GDP of approximately $180 billion, this equates to 0.4% of GDP.
- In Germany, the cost of grid reinforcement to mitigate harmonic issues has been estimated to be €2.5 billion or 1% of GDP.
- In Spain, a combination of the above has been estimated to have a cost of €6.5 billion or 1.2% of GDP.

The main issue with the above is that these are not real costs; rather, they are the hypothetical maximum costs associated with mitigating harmonics economically should that ever be implemented. The earlier four case studies are also cited in Ref. [7] and appear to be drawn from Ref. [6].

The study in Ref. [8] states that the harmonic losses in induction motors are about 0.3% of the rating. The commercial customer loss due to harmonic losses in neutral conductors is stated to be 1% of total direct costs. It is stated that this leads to a cost of $70 million for Australian customers. These figures are based on a series of assumptions that are not fully detailed in the study.

The study in Ref. [6] uses the results of measurement surveys to estimate the cost of PQ for European electricity supply networks. Whilst the study deals with a range of PQ disturbances, cost estimates are only given for harmonics. As a summary, it is stated that the costs of harmonics can be counted in the hundreds of billions of dollars worldwide.

The study in Ref. [9] states that 0.15%–0.2% of the total losses in network components are due to the existing level of harmonics.

In Ref. [5] a case study of a building containing 240 computers is cited which had a cost due to harmonic losses of $2101 per year. For utilities it is stated that transformers and generators are typically derated by 25% to accommodate harmonic distortion. Whilst this may seem like an excessively large derating factor, even a much smaller derating would suggest that if harmonic distortion is improved then there is potential to defer network augmentation expenditure by releasing network capacity.

4.5.2.2 Economic impact on network components

For network components, literature related to qualitative methods of calculating costs due to harmonics generally deals with calculation of the additional losses or premature ageing of equipment due to the presence of the harmonics. For example, in Ref. [10], for deterministic evaluation of the cost of harmonics, the following are presented as the main effects of voltage and current distortion on components and equipment:

- Additional losses
- Premature ageing
- Maloperation

Methods of calculating the costs associated with each of the above will be discussed in this section. However, the methods are data intensive and may be difficult to apply in practice. For example, in Ref. [11], for the effect of harmonics on transformers it is stated that the presence of harmonics increases core, copper and stray flux losses. The following equation shown is sourced from Ref. [12] and is stated to give the total load losses of a transformer when harmonics are present in the network.

It is clear that this equation requires a significant amount of data which might not be readily available to the network operator.

$$P_T = P_{CU} \times \left(\frac{I_L}{I_1}\right)^2 + P_{WEL} \times \left(\sum_n^n \left(\frac{I_n}{I_1}\right)^2 \cdot n^2\right) + (P_{CE1} + P_{SE1}) \times \left(\sum_n^n \left(\frac{I_n}{I_1}\right)^2 \cdot n^{0.8}\right)$$

where

- P_T is the power losses of a transformer due to harmonics
- P_{CU} is the total copper losses
- P_{WE} is the eddy current loss at 50 Hz at full load
- P_{CE1} is the additional eddy current losses at 50 Hz at full load
- P_{SE1} is the stray losses on construction parts at 50 Hz at full load
- I_n is the rms current (per unit) at harmonic n
- I_L is the value of the total rms load current
- I_1 is the fundamental component of the load current (per unit)
- n is the harmonic order

The other concerns stated for transformers in the presence of harmonics are overheating due to triplen harmonics circulating in delta windings and overheating due to increased loading. It is stated that the loss of lifetime of a transformer can be calculated based on hot spot temperature. An equation from Ref. [13] which is reproduced following show the calculation of relative ageing rate (V) as a function of hot spot temperature (note $V=1$ for a hot spot temperature of 98°C, an ambient temperature of 20°C and a hot spot temperature rise of 78°C).

$$V = \frac{\text{Ageing rate at } \theta_{hs}\text{°C}}{\text{Ageing rate at 98°C}} = 2^{(\theta_h - 98)/6}$$

where θ_{hs} is the hot spot temperature in degrees Celsius (°C).

For cables, it is stated that harmonic currents have two main effects:

- Additional ohmic losses (I^2R losses) in the line and neutral conductors of the cables due to the increased rms value of current when harmonics are present. This causes increased operating temperature.
- Generation of harmonic voltages due to interaction between harmonic current and network impedance. Harmonic voltages increase the dielectric stress on the cable and reduce lifespan.

The study presented in Ref. [14] provides equations for the heat generated in a cable consisting of '*m*' conductors and a harmonic component '*n*' and the loss of life caused by thermal degradation. These equations are reproduced as follows:

$$Q(m) = \sum m \cdot I_2^2 \cdot r_{ac}(n)$$

where

- $Q(m)$ is the heat generated in a cable per unit length
- M is the number of conductors in a cable
- $R_{ac}(n)$ is the conductor resistance for the nth harmonic per unit length
- I_n is the effective or rms value of the nth harmonic current

$$\rho = \rho_{rat} \cdot e^{-\left(\frac{E}{K}\right)\frac{\Delta\theta}{\theta_{\text{rat}}(\theta_{\text{rat}}+\Delta\theta)}}$$

where

- ρ is the lifetime referred to $\theta = \theta_{\text{rat}} + \Delta\theta$
- ρ_{rat} is the lifetime referred to $\theta = \theta_{\text{rat}}$
- $\Delta\theta$ is the temperature rise in relation θ_{rat} in Celsius
- θ_{rat} is the cable rated temperature in Kelvin
- K is the Boltzmann constant
- E is the materials activation energy

4.6 Management of harmonic distortion in electricity supply networks

Effective management of harmonic distortion in electricity supply networks requires coordination between the following:

- Emissions from equipment connected to the network
- The immunity/susceptibility of equipment connected to the network
- The level of distortion present or allowed on the network

As indicated by the above, management of harmonic distortion involves shared responsibilities between network users, equipment manufacturers/ suppliers and network operators. This section discusses the methods used to ensure that harmonic distortion on electricity supply networks remain below levels at which equipment will operate as expected and will not be damaged. Whilst there are many methods applied worldwide for management of harmonics in electricity networks, for the purposes of this section, emphasis is placed on the techniques prescribed by the IEC and the IEEE.

As a general principle, the responsibility for maintaining appropriate levels of harmonic voltage magnitudes is the responsibility of the network operators whilst the customer (noting that for the purposes of harmonics management a customer may be considered to be either a load or a generator) is responsible for the current emissions from their installation. The following passage from IEEE 519 [15] provides a succinct description of the shared responsibilities for management of harmonic distortion:

> *Harmonic voltage distortion limits are provided to reduce the potential negative effects on user and system equipment. Maintaining harmonic voltages below these levels necessitates that:*
>
> - *All users limit their harmonic current emissions to reasonable values determined in an equitable manner based on the inherent ownership stake each user has in the supply system.*
> - *Each system owner or operator takes action to decrease voltage distortion levels by modifying the supply system impedance characteristics as necessary [16].*

The relationships between network performance (i.e. voltage distortion magnitudes) and equipment immunity can be visualised by the diagram shown in Fig. 4.21.

In general terms, management of harmonics basically follows the philosophy detailed in Section 1.3 of Chapter 1. This philosophy is presented graphically as follows (Fig. 4.22).

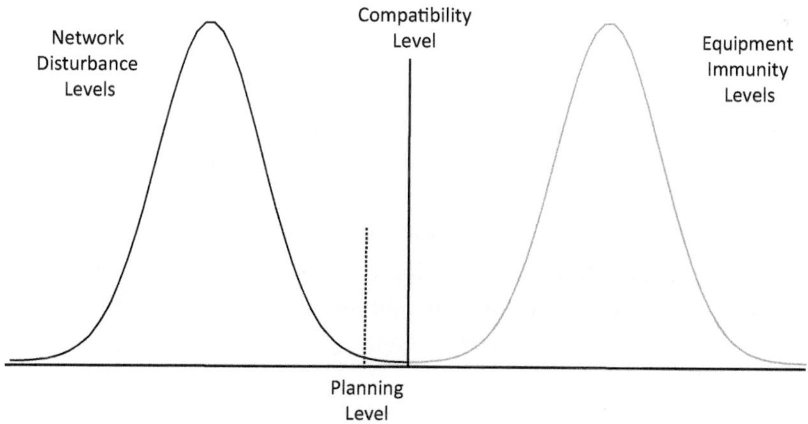

Fig. 4.21 Visualisation of power quality management philosophy which applies to management of harmonics.

Fig. 4.22 Flowchart for implementation of a harmonics management philosophy.

4.6.1 Compatibility levels

The IEC defines compatibility level as "the specified electromagnetic disturbance level used as a reference level for co-ordination in the setting of emission and immunity limits". The compatibility level is effectively the highest level of any disturbance magnitude that should be allowed to be present in the electricity supply network. Equipment manufacturers should use compatibility levels to design equipment to ensure that the equipment immunity level is greater than the compatibility level, thus ensuring that equipment operate as expected in the presence of power quality disturbance levels likely to be present in the supply voltage.

The IEC and IEEE have manifestly different approaches to defining voltage harmonic compatibility levels. For the IEC, there are different compatibility limits for odd, even and triplen harmonic orders and, as a general principle, the compatibility limit magnitude decreases as harmonic order increases. Fig. 4.23 shows the IEC compatibility limits for individual harmonic orders (magnitudes are in percentage of the fundamental voltage). The compatibility level for THD is 8%. It can be seen in Fig. 4.23 that higher magnitudes are allowed for odd order harmonics compared to even or triplen orders. There is some conjecture with respect to how this 'profile' of levels came about, as there does not appear to be any strong technical basis for allowable magnitudes decreasing as harmonic order increases. There is some evidence (including in IEC 61000-1-4 ED2) to suggest that these limits were determined based on measurements of harmonic magnitudes on existing power systems. This hypothesis is supported by the fact that the profile would match the expected emissions of the equipment (single phase and 6-pulse diode bridge rectifiers) connected to electricity supply networks at the time when the compatibility levels were first prescribed.

The IEEE takes an altogether different approach to compatibility levels. In fact, the IEEE does not specifically refer to a compatibility level (or planning levels which are discussed in the next section) but rather specifies voltage distortion limits. In contrast to the IEC, IEEE voltage distortion limits are at a flat value (i.e. the same value regardless of harmonic order) expressed

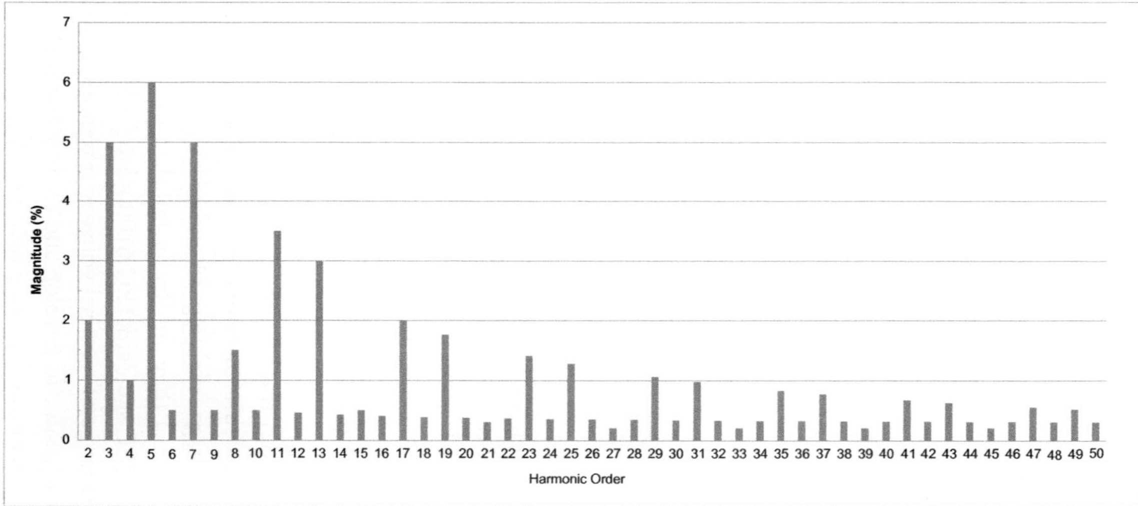

Fig. 4.23 IEC voltage harmonic compatibility levels from IEC 61000-2-2 [17].

Table 4.3 IEEE harmonic voltage distortion limits.

Bus voltage V at PCC	Individual harmonic (%) $h \leq 50$	Total harmonic distortion THD (%)
$V \leq 1.0\,kV$	5.0	8.0
$1\,kV < V \leq 69\,kV$	3.0	5.0
$69\,kV < V \leq 161\,kV$	1.5	2.5
$161\,kV < V$	1	1.5[a]

[a]High-voltage systems are allowed to have up to 2.0% THD where the cause is an HVDC terminal whose effects are found to be attenuated at points in the network where future users may be connected.

as a percentage of the rated power frequency voltage for all voltage harmonic orders as well as THD. IEEE harmonic voltage distortion limits are reproduced in Table 4.3.

4.6.2 Planning levels

Once compatibility levels have been established, using the IEC approach, electricity network planning methodologies must be implemented to ensure that power quality disturbance magnitudes present in electricity networks do not exceed the compatibility level. This generally results in another set of limits, termed planning levels, whose magnitudes are below compatibility levels. The IEC defines planning level as "level of a particular disturbance in a particular environment, adopted as a reference value for the limits to be set for the emissions from the installations in a particular system, in order to co-ordinate those limits with all the limits adopted for equipment and installations intended to be connected to the power supply system". The planning level effectively identifies the maximum disturbance magnitude that should be present at a particular network location. Planning levels are specified at each nominal voltage level (e.g. 11 and 33 kV) and, as stated before, values are selected such that they are always less than the compatibility level. Whilst the relevant IEC standard, IEC 61000-3-6 [18], provides indicative planning levels for a range of nominal voltages, it is ultimately the responsibility of the network operator to develop and apply planning levels for their network as part of the harmonic management planning process. As detailed before, the IEEE does not specify planning limits or have the same compatibility/planning limit philosophy as the IEC. Rather the harmonic voltage distortion limits presented in Table 4.3 apply.

4.6.3 Emission levels

As has been discussed previously in this chapter, voltage harmonics are caused by interaction between harmonic current and network impedance. As such, harmonic voltage magnitudes can be controlled through limiting the magnitude of harmonic current flowing in the network. The allowable magnitude of the harmonic current that may be 'emitted' by any piece of equipment and/or installation is termed as emission level. Emission level is defined by the IEC as "level of a given electromagnetic disturbance emitted from a particular device, equipment, system or disturbing installation as a whole, assessed and measured in a specified manner" [18].

For large installations connecting to medium voltage networks and above, it is typical to undertake an emission allocation process in order to limit the overall contribution that any particular installation (load or generator) may make to overall power quality disturbance levels through their power quality emissions. Both the IEC and IEEE have an approach to emission allocation; however, the methodology applied is very different in each case.

4.6.4 Management of harmonic distortion in low voltage networks

In low voltage networks, management of harmonic distortion is generally achieved by means of a combination of compatibility limits which provide an indication of the maximum magnitude of harmonic voltage that should be allowed on the supply network and emission standards for individual items of equipment which limit the magnitude of current harmonic emissions. Although the IEC has a methodology for allocation of emission limits to low voltage installations, IEC 61000-3-14, it is unusual for this technique to be applied in practice.

4.6.5 Management of harmonic distortion in medium and high voltage networks

In medium and high voltage networks, both the IEC and IEEE prescribe an emission allocation process which aims to control voltage distortion magnitudes. Regardless of the emission allocation technique, the overarching principle is that each customer connecting to the network, whether it be a load or a generator, is allocated an emission level (i.e. maximum magnitude) for each harmonic order. This may also be thought of as apportioning the harmonic absorption capacity of the network. Once allocated an emission limit, it is the responsibility of the customer to ensure that the allocated emission level can be met.

One of the key challenges of emission allocation is to ensure that the full capacity of the network to absorb harmonics is utilised (i.e. when all load has been connected voltage distortion should be close to planning levels or limits) whilst in turn ensuring that waveform distortion planning levels or limits are not exceeded. As such, there is always tension between over- and underallocating emissions. A conservative approach to allocation, i.e. allocating emission levels that may be lower than necessary, can lead to significant difficulty for connected customers including the need to instal costly harmonic current mitigation devices. Conversely, overallocation can lead to excessive harmonic voltage distortion levels leading to undesirable impacts on both customer and network equipment.

A comprehensive analysis of harmonic emission allocation processes in use around the world indicates that whilst there exists a large range of allocation processes and methodologies, in high level terms the majority can be grouped into two specific categories. For the purposes of this chapter, these two allocation methodologies have been given the terms:

- Fixed harmonic allocation methodologies
- Network forecast methodologies

A fundamental difference exists between the two allocation methodologies. Fixed allocation methods allocate emission limits to connecting installations independent of the state of the network (existing or forecast). The fundamental philosophy being that the assumptions made during the development of the allocation method will result in the harmonic voltage magnitudes being below or just reaching required levels if all customers are meeting their allotted emission allocations. In the event that the system exceeds harmonic voltage limits, it is the responsibility of the network operator to implement mitigation. In contrast, network forecast methods allocate harmonic emission levels on a pro rata basis based on the agreed power of the connection and the total agreed power of all connections (existing and future) within the system. This approach attempts to implement engineering judgement with respect to future states to ensure planning levels are never exceeded if all customers are meeting their allocated emission limits.

4.6.6 Fixed harmonic emission allocation methodologies

Fixed emission allocation methods allocate a predetermined percentage value of harmonic voltage or current based on simple details related to the connection request. The simplest approach to fixed allocation can be found in Refs. [19,20] which are applied in Finland and Brazil, respectively. A discrete harmonic voltage is allocated to the connecting customer based solely on the connecting voltage level of the installation, i.e. allocation is independent of the agreed power and/or Short Circuit Ratio (SCR) of the installation. Fixed allocation methodologies such as those found in Refs. [21–23] (IEEE, Hydro-Quebec and France, respectively) allocate a discrete emission limit to the installation based on the SCR at the PCC. This is done via a lookup table an example of which, from IEEE 519, is presented in Table 4.4 [21].

The method used in Canada [22] uses a similar table; however, linear interpolation is used to scale the harmonic emission limit based on the actual SCR of the installation. Some of the fixed allocation methods use scaling factors and equations to define the final harmonic emission limit [23,24]. For example, the process in Ref. [24] uses the following equation to allocate harmonic current emission limits to connecting installations.

$$\frac{I_v}{I_A} \leq \frac{p_v}{1000} \cdot \sqrt{\frac{S_{kV}}{S_A}}$$

where

- I_v is the harmonic current in A
- I_A is the nominal operating current of the installation in A
- p_v is a proportionality factor based on the compatibility levels for the harmonic order of interest (h) and the impact of harmonic interaction across different voltage levels, given in Ref. [24]
- S_{kV} is the short-circuit power of the network at the PCC
- S_A is the nominal power of the installation

The approach in Ref. [24] which is applied in Europe removes the need for a lookup table as the resulting harmonic emission allocation for all orders is calculated. However, the allocation is still considered as fixed, as the variables of the previous equation are not impacted by the consideration of future forecasts, but instead use fixed scaling factors such as p_v. Conservative assumptions and scaling factors are introduced to allow for uncertainty, for example, diversity of emissions, varying network characteristics and resonance.

Fixed emission allocation methodologies prioritise simplicity of application over detailed assessment of network characteristics or future connections. The process by which this is achieved is by developing a number of conservative assumptions about the range of networks to which they

Table 4.4 IEEE 519 harmonic current allocation for voltages $0.12\,kV \le V_{nominal} \le 69\,kV$ [21].

Maximum harmonic current distortion in per cent of I_L						
Individual harmonic order (odd harmonics)						
I_{SC}/I_L	$3 \le h < 11$	$11 \le h < 17$	$17 \le h < 23$	$23 \le h < 35$	$35 \le h \le 50$	TDD
<20	4.0	2.0	1.5	0.6	0.3	5.0
$20 < 50$	7.0	3.5	2.5	1.0	0.5	8.0
$50 < 100$	10.0	4.5	4.0	1.5	0.7	12.0
$100 < 1000$	12.0	5.5	5.0	2.0	1.0	15.0
>1000	15.0	7.0	6.0	2.5	1.4	20.0

are being applied. These assumptions are implemented to account for impacts such as resonance or reduced diversity of emission. However, these assumptions are also capable of leading to overly conservative allocation limits. Further, if the network parameters to which the process is being applied significantly differ from the characteristics assumed in the development of the allocation process, harmonic voltage limits may be exceeded. Fixed allocation method advantages and disadvantages that have been identified are as follows:

Advantages:

- Simplicity of implementation (i.e. basic calculations)
- Minimal input data requirements

Disadvantages:

- Underlying assumptions not designed to be applied to all network types
- Assumptions can be shown to be overly conservative in general but also capable of exceeding planning levels in some cases
- Increased uncertainty related to renewable energy generation exacerbates issues related to conservative assumptions

4.6.7 Network forecast methodologies

The most frequently applied network forecast emission allocation methodology is the IEC approach [18]. This allocation methodology requires the network operator to define harmonic voltage planning levels (as discussed before), described as internal objectives for the operator. As customers begin to connect to the network, they are allocated a portion of the planning level based on the relative size of the proposed installation (S_i) and the peak forecast of connected loads at the point of common coupling (PCC) allowing for diversity (S_t).

The IEC emission allocation methodology consists of a three-stage approach as follows:

- *Stage 1*: If an installation is sufficiently small relative to the strength of the network at the PCC (i.e. $SCR \geq 500$), connection may be accepted without any further consideration.
- *Stage 2*: If the installation does not meet the requirement for Stage 1, harmonic emission limits are allocated using the following equation, assuming all loads are connecting to the same PCC.

$$E_{Uhi} = G_{h,MV} \cdot \left(\frac{S_i}{S_t}\right)^{1/\alpha}$$

where

- E_{Uhi} is the allocated harmonic voltage provided by the network operator to the connecting installation
- $G_{h,\ MV}$ is the global contribution of harmonic voltage emissions allowable at the PCC from MV installations
- S_i is the agreed power of the installation under consideration
- S_t is the aggregate agreed power of all existing and projected connected installations
- α is the summation exponent to allow for diversity of emissions (both in time and phase angle) between harmonic sources. Relevant values are given in Ref. [18]

The previous equation allocates a harmonic voltage to the installation which can also be converted into a current using the harmonic impedance of the network at the PCC. In the scenario that:

- Aggregated agreed power of all connected installations is equal to S_t, i.e. the network is fully loaded
- All installations have received an allocation
- All installations are emitting their allocated limits

There exists some contention with respect to the definition and therefore assessment of compliance for E_{Uhi} as the actual harmonic voltage due to the connection of the installation is entirely dependent on the level of harmonic voltage distortion and operating state of the network [25]. For the purposes of this chapter, E_{Uhi} may be defined as the decremental harmonic voltage magnitude when the installation is disconnected from the network with background harmonic magnitude at the planning level, i.e. from a compliance assessment standpoint:

$$E_{Uhi} = L_h - V_s$$

where

- L_h is the harmonic voltage planning level
- V_s is the harmonic voltage due to all other plants other than i

- *Stage 3 (acceptance of higher emission levels)*: If an installation is unable to meet the emission levels allocated using the Stage 2 methodology, under the auspices of AS/NZS 61000.3.6, a call for mitigation of emissions is only one of the options available to the network operator. Another option is to allow short-term acceptance of a connection under the provisions of a Stage 3 allocation. These provisions allow the installation to be connected in the short term with a longer time period available to finalise mitigation measures. Whilst Stage 3 connection is provided as an option in IEC 61000-3-6 there is little guidance as to when or how it should be applied nor for how long.

4.6.7.1 Preconnection compliance assessment

In order to determine whether an installation can meet allocated emission levels, a study is undertaken at the planning stage. This study is often termed as preconnection compliance assessment. Preconnection compliance assessment for loads and generators is undertaken by means of numerical modelling, most often implemented using computer software. In simple terms, the current emission of the installation is injected into the impedance of the network and the resulting harmonic distortion magnitudes are compared to the allocated emission levels. To be effective and accurate, the preconnection compliance assessment process requires a good understanding of the network impedance and load/generator harmonic emission characteristics (emission spectrum), both of which may be required to consider hundreds of operational scenarios. The accuracy of the outcomes of the studies is directly impacted by the accuracy of the data available.

As a general principle, the meaning of 'harmonic voltage emission' in the preconnection compliance assessment process requires a more prescriptive definition. At present there is scope for various interpretations, and this can lead to conflict between NSP and customer. One commonly used interpretation leads to small customers being asked to meet unnecessarily onerous assessment conditions.

The following have been identified as areas of uncertainty with respect to preconnection compliance assessment:

1. Representation of network impedance
2. Accuracy of harmonic models for power electronic devices and/or harmonic emission spectra provided by equipment manufacturers/suppliers
3. Application of AS/NZS 61000.3.6 Stage 3 which, in its present form, provides an option for conditional connection but little further detail

Each of the above is investigated in detail as follows.

Modelling of network impedance

The frequency-dependent network impedance values used for preconnection compliance studies are not consistent across all network operators and present standards that lack prescriptive detail with respect to how network impedance should be determined. Until recently it was common for the network impedance to be represented as a single value corresponding to the most likely network operating scenario. In many cases, this single value was simply the fault level at the point of connection and was assumed to be inductive. Particularly at transmission level, it is now becoming more common to represent network impedance as an impedance polygon (one polygon for each harmonic order). An impedance polygon is generated using computer software to derive the impedance at the point of connection for a wide range of network operating scenarios. Each impedance value can be plotted onto an impedance plane (resistance/reactance). The polygon is then created by drawing lines to encapsulate the impedance values. A polygon is often characterised by the way in which these lines are drawn. For example, a 10-point polygon is created by joining 10 of the impedance values. An example of a 10-point harmonic impedance polygon is shown in Fig. 4.24.

Both representations of impedance, single value or impedance polygon present uncertainties that should be considered when undertaking precompliance assessment. With respect to supply of a single impedance value, the uncertainty is reasonably straightforward; the single impedance value may not represent scenarios at which the network is operated for extended periods of time. Under- or overestimation of the single impedance value may either result in underutilisation of the harmonic absorption capacity of the network, i.e. the ability of the network to absorb harmonic current emissions without significant increase of harmonic voltage distortion levels or noncompliance at the preconnection stage that will not be observed once the installation is commissioned. In addition, simple scaling of the single fundamental frequency impedance value to account for impedance changes as harmonic order increases will not allow assessment of any resonance conditions where impedance can be significantly lower or higher than expected.

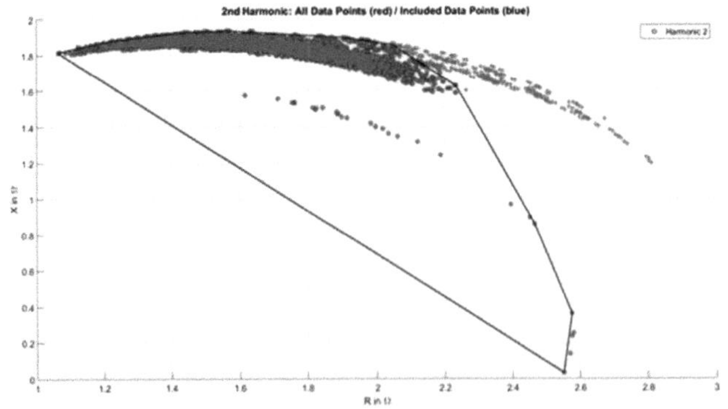

Fig. 4.24 Example of a harmonic impedance polygon.

Harmonic impedance polygons remove some of the uncertainty related to single impedance values as the impedance polygon can include many network operating scenarios. However, a range of issues have been encountered related to the generation of harmonic impedance polygons including:

- A general lack of understanding or information with respect to the most likely operating scenarios for networks, especially concerning:
  - Service status and appropriate modelling data regarding reactive plants (i.e. power factor correction capacitor banks)
  - Service status of $n-1$ components (e.g. redundancy of transformers and transmission/distribution lines)
  - Understanding of harmonic damping due to network loading
- A general lack of guidance with respect to the extent to which the network should be modelled. For example, is modelling the network several nodes away from the connection point of an installation suitable or is a more comprehensive model required? It is common for NSPs to require modelling of very large networks, increasing model complexity and associated time to develop and implement with uncertain benefit to accuracy.
- Harmonic impedance polygons represent the network at the time when they are generated. In some cases, the polygons may also omit parts of the network, e.g. harmonic filters or committed installations (i.e. installations that are yet to physically connect to the system but have the appropriate connection agreements in place). The harmonic polygons do not reflect future network conditions and as such, installations may be assessed as being compliant at the planning stage only to be assessed as noncompliant postconnection.

A prescriptive methodology for the generation of harmonic impedance polygons does not presently exist. The present practice of polygon generation generally involves considering a large number of different network operating scenarios for large networks with little or no weighting given to the likelihood of any particular scenario occurring. In many cases, the least likely network operating scenario is concurrent with the largest impedance. Including these unlikely, high impedance, network operating scenarios increases the likelihood of harmonic emission mitigation being required by the proponent. Consideration must also be given to the fact that compliance is generally assessed based on 95th percentile values of measurements. That being the case, is it reasonable to include impedance values that will not likely exist for more than 5% of the time? A number of network operators have taken the fact that compliance is based on 95th percentile values into account by omitting the largest 5% of impedances in generated impedance polygons. However, it has been shown that the most likely network operating scenario can occasionally exist within the largest 5% of impedances and as such may be arbitrarily omitted. Further, there is at present no consideration for how future network harmonic impedance may be accounted for. Subsequently, it is possible for existing plants to become noncompliant due to the connection of installations in the future.

Modelling process for power electronic connections

Accurate modelling of the emission spectra of power electronic devices is a key requirement for accurate preconnection compliance assessment. This modelling has become significantly more important as the number of renewable energy connections interfaced to the network through inverter-based resources (IRBs) has been increasing rapidly. A significant volume of research has been completed that identifies the impact of operational conditions on the harmonic emissions of a power electronic device [26–28]. Impacts of both internal and external conditions, e.g. network operating state and device operating power, are capable of significantly varying the emissions of a device; however, these variations are often ignored in preconnection compliance studies in favour of a single set of harmonic emission levels generated based on unknown network characteristics and with an uncertain/nonstandardised measurement methodology. OEMs of large inverters have specifically identified characterisation of inverter emissions as being highly challenging especially as the rating of the inverter increases. Operational conditions of networks that have been shown to vary the harmonic current emissions of power electronic devices include, but are not limited to:

- Magnitude and order of background harmonic voltages
- Magnitude of fundamental voltage
- Magnitude of network unbalance

Operational conditions of power electronic devices themselves that have been shown to vary the harmonic current emission include, but are not limited to:

- Inverter operational set point, i.e. active and reactive power output
- Operational state of the device controller, i.e. Constant P/Q/V, Volt/Var, Volt/Watt

The variables discussed before are considered in IEC TR 61400.21.3 [29], which is the international technical report currently being used to determine the emission characteristics of inverters by means of laboratory testing, time domain simulation, converter impedance calculation or a combination of some or all of these approaches. This technical report provides general guidance in the development of harmonic models for wind turbines only, although it is suggested in Ref. [28] that the process be extended to all devices using power electronic converters. At present, the onus is placed on the device manufacturer to decide the conditions for which the emission models are developed and, in general, only a single set of worst-case emission spectra is provided which does not necessarily correspond to a single power output level, e.g. maximum power output. Underestimating emissions at the preconnection stage may result in higher than expected impact postconnection whilst overestimating emissions at the preconnection stage may result in preconnection noncompliance requiring harmonic mitigation for the installation that is ill designed or may not be required once the installation is connected. Similar to other technical reports, there lacks a prescriptive method in Ref. [29] (or other technical documents) which can be followed to achieve a standardised outcome, in this case, a harmonic model for the purposes of conducting preconnection compliance assessment.

In addition to the above, the assumption that inverter-based resources behave as current sources across all harmonic orders may be too simplistic and lead to incorrect deductions from inverter laboratory measurements and unnecessarily conservative assessments of preconnection compliance. Alternative models need to be robust, accurate and implementable. Available modelling approaches include time domain modelling and frequency domain models, such as frequency-dependent Norton impedances although difficulties have been identified with both approaches. Time domain modelling is the most complex and therefore computationally intensive approach to harmonic modelling. Time constraints and complexity of development reduce the practicality of this approach and increase the time required to undertake the assessment. Issues have also been identified with relation to the accuracy of models for harmonic emissions studies. Frequency domain modelling reduces the complexity of modelling by representing harmonic sources as simple linear components such as Norton equivalent circuits. However, the validity of this simplification has been shown to be affected by a range of impacts as previously discussed. Nonrepresentative simplified models are also capable of resulting in improper harmonic filter design.

4.6.8 Impact of renewable energy generation on emission allocation

The increasing penetration levels of large renewable energy generators (REG) change the traditional load-related paradigm with respect to allocation of harmonic emissions in high voltage networks. The reason for this is that there will likely be many more REG connections than is the case for high voltage loads. Further, REG may be able to be connected even if the load capacity of the network has been reached. As such, in the presence of REG, a review of the previous approaches to emission allocation is warranted. If the fundamental processes of the traditional approaches are retained, it is highly likely that more conservative assumptions will be required to ensure harmonic voltage emissions remain below defined planning levels for all likely scenarios. However, increasing the conservatism of the present allocation methodologies is likely to lead to a decrease in the efficient use of the absorption capacity of many networks to which the harmonic emission allocation methodology is to be applied. Many of the fixed allocation methodologies also apply restrictions on generating plant, including REG, based on the principle that the characteristics of all generating installations emissions should be similar to those of large synchronous generators, i.e. be characterised as a low harmonic impedance with minimal harmonic current injection. It is well known that these principles are not applicable for REG such as solar PV and wind that interface to the network by means of power electronic converter (inverter). These installations are a source of harmonic emissions, similar to power electronic loads.

One of the key uncertainties that exists with respect to deployment of REG is the connection rate (i.e. how much REG is to be connected and over what time frame). During the planning phase of a new or upgraded network, for the purposes of emission allocation, it is necessary to estimate the peak load forecast (S_t). Estimation of this value is subject to uncertainty and can result in significant impacts when estimated peak loads are not realised. This uncertainty with respect to estimates of peak load can have significant implications for management of harmonic emissions. For example, a network planner may expect that a particular area of a network is likely to connect large volumes of REG and plan accordingly. However, the final level of penetration of REG may be far more (or less) than expected, resulting in the harmonic absorption capabilities of the network being either over or underutilised.

A practical approach to mitigating the uncertainty related to REG connection is for the network operator to allow for some level of REG to be incorporated into S_t based on certain, probable and uncertain connections. A comparative infographic for this approach is shown in Fig. 4.25 where:

(a) represents the present approach, that is, REG is not planned for and thus the system will reach harmonic voltage planning levels prior to the thermal capacity of the system being fully utilised. This results in underutilisation of the system harmonic absorption capacity.

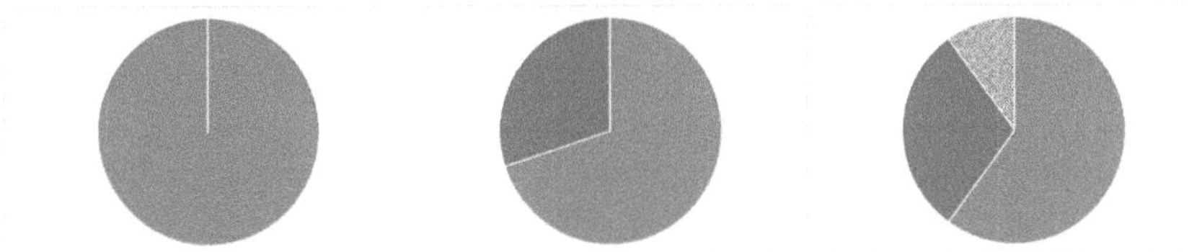

Fig. 4.25 Comparison of methods that could be used to by a network operator to account for REG in estimation of peak load *blue (dark gray* in print version)—peak forecast load, *orange* (*gray* in print version)—certain REG, *red* (*light gray* in print version) hatch—uncertain REG.

(b) represents a situation where the network operator accounts for REG connections that are certainly ahead of time and thus apportions this into the harmonic management process using:

$$S_t = S_c + S_G$$

where

- S_c is the sum of the agreed power for all expected standard consuming installations
- S_G is the sum of the agreed power for all expected generator installations

Whilst this is the ideal scenario, given the uncertainty of generator connections and expected penetration levels for some networks it is unlikely to be the case in practice.

(c) represents perhaps the most likely scenario. Here there is a level of uncertain REG connections to be accounted for, providing a buffer for uncertain generators although reducing the total level of allocated distortion. The way in which uncertain REG installations are apportioned can vary depending on the level of uncertainty.

A number of approaches to mitigate the impact of the uncertainty of REG penetration levels on emission allocation have been suggested in recent years. One of the more basic considerations was included in the 2008 revision of IEC 61000-3-6, where it is suggested that the agreed power of REG is to be included into the value of S_t; however, it also states that a more detailed consideration is required to better understand the contribution of the generator to both the system supply capacity and fault level. The idea of including REG agreed power in S_t is extended in Refs. [30,31] which suggests an updated definition of S_t to include expected REG, i.e.

$$S_t = \lambda_c S_c + \lambda_p S_p$$

where

- Subscript c identifies the variables related to consuming installations (conventional loads)
- Subscript p identifies the variables related to generating installations (generators)
- λ_c is a scaling factor applied to standard consuming loads
- λ_p is a scaling factor applied to generating connections

This approach can augment the IEC emission allocation methodology. However, the values for each parameter are network specific, so incorrect assumptions are capable of excessively strict allocations to connections that eventuate in favour for those that do not.

A related approach has been presented in Ref. [32] specifically for the allocation of offshore wind farms in the Netherlands and Germany. Investigation of the defined harmonic emission limits provided identifies that the limits have been halved in comparison to the emission limits that would be allocated using the traditional IEC method. Such a reduction in allocated emission limits is akin to the use of the previous equation with:

$$S_c = S_p$$

and

$$\lambda_c = \lambda_p = 1$$

While this approach may be acceptable in unique cases, the scenario in which it is applied in Ref. [32] is for an offshore platform, i.e. a system with minimal consuming load. Therefore the level of generated power is likely to far exceed the level of consuming power, i.e. $S_c \ll S_p$, needlessly limiting the harmonic emissions of the generators.

Until recently, the IEEE emission allocation methodology addressed REG within IEEE standard 519 by stating that generating installations are to have minimal impact on the distortion of the network and thus are limited to the lowest allowable harmonic emission limit, regardless of SCR. However, the 2022 revision to Standard 519 refers emission allocation for installations that are predominately REG (noting that Standard 519 remains applicable for installations with a mix of load and REG) to Standard 1547 (IEEE Standard for Interconnection and Interoperability of Distributed Energy Resources with Associated Electric Power Systems Interfaces) and Standard 2800 (IEEE Standard for Interconnection and Interoperability of Inverter-Based Resources (IBRs) Interconnecting with Associated Transmission Electric Power Systems). Standard 1547 applies to distribution systems whilst 2800 applies to transmission and subtransmission systems. Both the standards provide fixed current distortion limits for REG interfacing to the distribution network by means of inverters.

4.6.8.1 Uncertainty related to previous harmonic allocations

Another difficulty related to the IEC approach to emission allocation is the uncertainty related to the methodology used for previous emission allocations. The practical implementation of the IEC approach is predicated on all installations within a network being allocated emission limits using the same methodology. However, historically, the IEC approach has been introduced to many existing networks, with existing customers, each potentially with a connection agreement that already defines their harmonic emission limits which may have been allocated using a different methodology to the latest IEC approach. Recent updates to existing allocation methodologies address this uncertainty in slightly different ways by implementing what is colloquially known as the 'headroom approach'. Using this approach, measurements of existing harmonic magnitudes are used to determine the level of distortion present in the network at a point in time and then allocate a portion of the remaining headroom (i.e. the difference between the planning level and the measured value) to the connecting installation. While implementation of the headroom approach is capable of addressing the uncertainty of previous allocations, networks implementing this approach are susceptible to further difficulties as described as follows.

Difficulties in the application of existing headroom approaches

The process in Ref. [33], which applies in the United Kingdom, is concerned with both calculating harmonic allocations and harmonic management in general. The approach is a three-stage process where a connecting installation must provide adequate evidence that the connecting equipment/installation is capable of meeting harmonic emissions compliance at one of the relevant stages. Progression through the stages increases the data required and complexity of the process. The three stages are defined as follows:

- Stage 1—Includes four substages and is relevant to LV connections only
- Stage 2—Includes three substages, relevant to MV connections and LV connections that fail Stage 1
- Stage 3—Implements the headroom approach for all connections failing to meet the requirements of Stage 1 and 2

Both Stages 1 and 2 do not allocate harmonic emission limits but instead make a number of conservative assumptions to allow simple studies for installations that are unlikely to have a significant impact on harmonic voltage magnitudes and/or network impedance. The installation is considered to have complied with requirements if any of the relevant studies suggest that the network will not exceed planning levels postconnection or if the connecting equipment meets a number of international standards related to harmonic.

Stage 3 allocation is required for two reasons:

- The background harmonic voltage levels are above 75% of the planning levels or
- The identified harmonic current emissions from the new installation cannot meet the requirements of Stage 1 and 2 (based on pessimistic assumptions regarding the network impedance).

In the event that the background harmonic voltage levels are approaching planning levels and a connection requires an emission allocation, this is completed by determining the available headroom at the PCC (or any bus of interest within the network) by using the following equation:

$$V_{h,HR,PCC} = \left(L_{h,MV}^{\alpha} - V_{h,BG,PCC}^{\alpha}\right)^{\frac{1}{\alpha}}$$

where

- L_{MV} is the planning level at the PCC
- $V_{h,\ BG,\ PCC}$ is the background harmonic voltage level at the PCC
- α is the diversity factor, as defined in Ref. [18]

This process is also completed for any bus of interest in the network and the minimum available headroom is apportioned to the connecting installation based on voltage level. If the nominal voltage is less than or equal to 132 kV, the installation is allocated 50% of the remaining headroom. If the nominal voltage is greater than 132 kV, the allocation is scaled based on the agreed power of the installation and the connecting voltage level.

Using the headroom approach, it can be shown that no installation is ever allocated the entirety of the remaining headroom and therefore the total of allocated harmonic emission levels should never exceed planning levels, provided all installations are meeting their allocation requirements. However, for networks operating at or below 132 kV, installations receive 50% of the remaining headroom regardless of the agreed power. This approach can be shown to be disproportionately biased towards early connections, omitting the objective of fair emission limit allocation as specified by the IEC methodology. This preponderance to allocate higher emissions to early

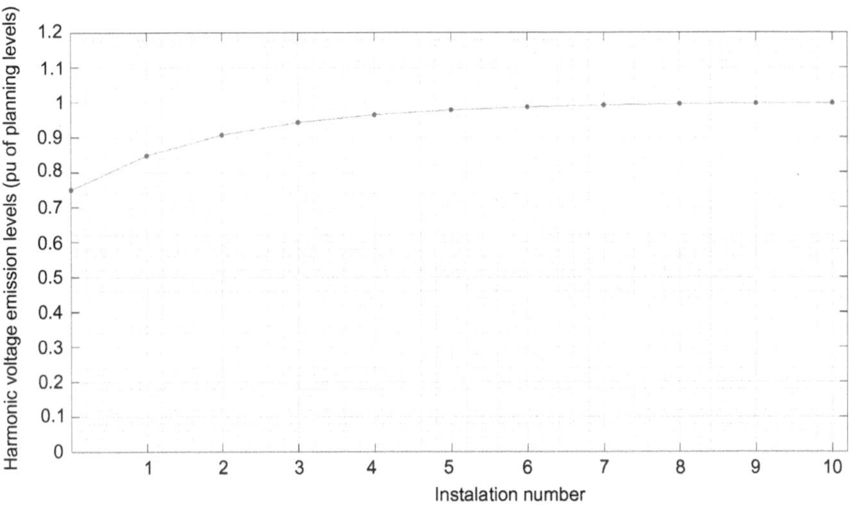

Fig. 4.26 Harmonic voltage growth profile using Headroom allocation approach showing early connections receiving higher emission allocation.

connections is demonstrated graphically in Fig. 4.26, in which the harmonic voltage profile of a network evolves over time due to continued connections.

As shown in Fig. 4.26, once the background harmonic emission level is at 75% of the planning level a new connection is allocated 50% of the remaining headroom, regardless of its agreed connection capacity. Due to allowance for diversity, the expected emissions from the new plant increase the background harmonic distortion to 85% of the planning level. The next installation receives an allocation of 50% of the remaining headroom and so on. Whilst the remaining headroom is never fully allocated, unrealistic restrictions are imposed on installations solely based on the point in time at which the connection request occurs. The process is also capable of being too lenient on earlier installations. For these reasons it can be shown that this harmonic allocation approach may be capable of efficient usage of the power system harmonic absorption capacity and ensuring that planning limits are not exceeded but is not capable of fair allocation across all customers, a key objective of the IEC methodology.

Recent alterations to the IEC allocation methodology made by the Irish TSO, EirGrid, detailed in Ref. [34], introduce a requirement for monitoring prior to harmonic emission limits being calculated. The first step is to determine the background harmonic voltage level at the PCC. The background harmonic voltage levels (V_{BG}) are then used to determine the available headroom at each harmonic frequency at the node under study using the following equation:

$$V_{\text{headroom}} = \left(L_h^{\alpha} - V_{h,BG}^{\alpha}\right)^{\frac{1}{\alpha}}$$

where

- V_{headroom} is the harmonic voltage headroom at the location of interest within the network
- L_h is the harmonic voltage planning level
- $V_{h,\ BG}$ is the measured background harmonic voltage distortion at the location of interest
- α is the summation exponent as defined in Ref. [35]

A portion (25%) of the available headroom ($V_{h,\ \text{remaining}}$) is retained as a conservative measure to ensure planning levels are not exceeded for any connecting installations when meeting their allocated levels. The remaining headroom available for allocation is therefore:

$$V_{h,\text{remaining}} = V_{\text{headroom}} \cdot (1 - 0.25)$$

Allocation of the harmonic voltage distortion limit is then completed by apportioning the remaining headroom available on a 'MW pro rata basis', i.e. following a modified version of the IEC Stage 2 allocation approach using the following equation:

$$E_{Uhi} = V_{h,\text{remaining}} \cdot \left(\frac{S_i}{S_t}\right)^{\frac{1}{\alpha}}$$

Whilst this approach is able to pragmatically allocate the remaining headroom and removes the difficulty introduced in the approach presented in Ref. [33], ambiguities in the description of the process are capable of resulting in disproportionately conservative harmonic emission limits. The explanation of the allocation method in Ref. [34] suggests that harmonic monitoring and calculation of remaining headroom (including planning margin allowance) is performed for every connecting installation. However, the division of available headroom suggests that the background harmonic voltage levels are measured once and the remaining headroom is apportioned to *n* customers, with the final allocated harmonic distortion limits reaching 0.75 pu of the low voltage planning level. The impact of this discrepancy can be shown by considering two scenarios:

1. Remaining headroom of the system is monitored and calculated for every connecting installation

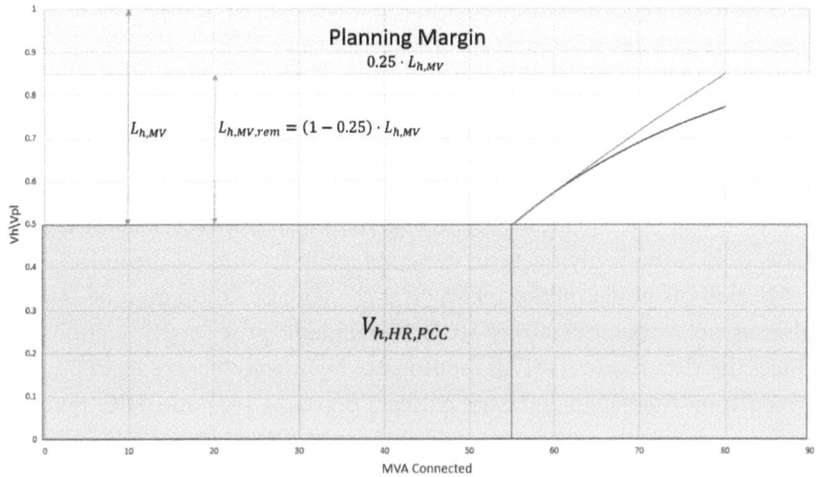

Fig. 4.27 EirGrid allocation example with existing harmonic sources, $S_{connected} = 55$ MVA [*yellow* (*light gray* in print version)], V_h/V_{PL}—Scenario 1 [*red* (*dark gray* in print version)], V_h/V_{PL}—Scenario 2 [*blue* (*gray* in print version)].

2. Headroom is calculated once and apportioned to the remaining installations

The result of both of the earlier scenarios is shown in Fig. 4.27 where it can be seen that Scenario 1 results in a conservative use of the network harmonic absorption capacity whilst Scenario 2 meets the expected planning level once all installations are connected. The conservative result for Scenario 1 is due to the repeated calculation of available headroom including planning margin for every connection, reducing the total allocated emission limits.

4.6.8.2 Uncertainty of REG connection location and volume

A number of proposed updates to the existing IEC emission allocation methodology have been published since its inception in 1996. These updates have been provided to address a number of issues and have had varying levels of impact on the allocation.

The revised methods presented in Refs. [36–39] define a process which reduces the disparity of allocated emissions for connections of similar agreed power connecting to positions on the network with significant differences in fault levels (e.g. sending/receiving ends of long feeders). The methods determine the position on the network that will reach planning levels first and implement a scaling factor to the equation and allocate emission limits on a pro rata basis to all connections. It is identified in Refs. [37,38,40,41] that this approach requires an unrealistic knowledge of future loading that exposes the process to significant impacts of uncertainty. This is addressed in Refs. [37,38] by providing a simplified process to determine the weakest feeder in the system. However, the method adopts the assumption that all connections are evenly distributed along the feeders. It can be shown that such an assumption is too general and can lead to incorrect calculations of where the system will reach planning levels first. This is an important factor as REG connections commonly occur in geographically remote locations [34,42].

Another proposed method, known as the voltage droop method [40,41], addresses network uncertainty with a pragmatic process that significantly reduces the data required. This method has been well received and is incorporated into Australian industry guiding practices [43] and IEC revisions which are currently underway [31]. However, there exists limitations to the method when the presence of REG is taken into account [43]. Further, the voltage droop method can be shown to consistently allocate more conservative limits than the standard IEC method [40]—already one of the more conservative allocation procedures based on a simple study presented in Ref. [44]. Although the impact of different network characteristics when comparing the voltage droop methodology with other allocation methodologies must be taken into account [44–46]. As such, widespread application of the voltage droop method may lead to disproportionately stringent emission limits being applied.

Whilst a significant amount of work has already been completed in relation to the practical application of the IEC emission allocation methodology, there remain a number of significant issues when considering the integration of REG. As identified, the proposed methods introduced to address the significant level of data required for strict application of the IEC method can in turn be too general, omitting considerable impacts of concentrated loads. Further concern related to the application of the voltage droop method, i.e. overly stringent limits being applied, and concern of REG have also been identified.

4.6.8.3 Uncertainty of future harmonic impedance of networks

Significant difficulties have been identified by network operators and regulators, with regard to the changing impedance of future electricity networks. For example, following the significant increase in renewable energy generators, Australia is expected to begin retiring its ageing fleet of synchronous generators. The Australian Energy Market Operator (AEMO) anticipates that 63% of existing synchronous generation will be lost from the Australian

National Electricity Market by 2040. This poses a difficulty as synchronous generators provide a low impedance path for harmonic currents, generally reducing voltage distortion.

Mitigation of harmonic distortion by means of connection of network strengthening devices such as synchronous condensers, which provide similar characteristics to synchronous generators, is expensive and only considered if the condenser is required to solve other issues, e.g. voltage stability or reactive power support. Another potential solution is the use of harmonic filters; however, continued connection of filters is capable of shifting resonant frequencies to problematic values and may ultimately amplify emissions of other installations.

Based on all of the above, it is clear that managing harmonic emissions in an environment of continuing REG penetration is likely to remain a challenge now and into the future.

4.6.9 Postconnection compliance assessment

Assessment of postconnection compliance with allocated emission levels by means of field measurements is another subject of importance as REG penetration increases. At present, there is no clear consensus with regard to the methodology for measurement-based compliance assessment despite there being much research over at least a decade [47–51]. Part of the problem has been incorrect or poor understanding of the concept of harmonic emission (or the definition of harmonic voltage emission). The following are key uncertainties and questions to be addressed in relation to postconnection compliance assessment:

- Customer's responsibility: Should it be defined in terms of voltage, current or some other concept? Is the concept of 'customer's share' rigorous and is it unambiguously measurable?
- Plant operating conditions: Solar and wind farm operation depends on uncontrollable and sometimes unforeseeable atmospheric conditions. What are the allowable conditions for reasonable compliance measurement?
- Power system operating conditions: Should checks be made as to the operational scenario(s) during the test period? Are all scenarios equally useful for assessing customer compliance?
- Measurement quantities: Large plant contains transformers giving a range of internal voltage levels and possibly shunt filters and capacitors. Should voltage and current measurements be taken only at the point of connection or at other voltage levels as well?

- Measurement duration: Can compliance be assessed over a day or is a longer period required. When seasonal effects are important in renewable generation, how should these be allowed for?
- Signal processing: Can assessment be based on raw data or some processed quantity? Are there tests that need to be implemented for the suitability of data for compliance determination?
- What are the ongoing installation requirements with respect to discharging compliance obligations?
- What impact do measurement transducers, particularly voltage transformers, have on compliance assessment and can recommendations with regard to appropriate measurement equipment be made?

4.7 Mitigation of harmonic distortion

As with mitigation for most power quality disturbances, there are many options that can be used to mitigate harmonic distortion. Solutions can be either at the consumer or at the network level. Sound engineering judgement which considers both the technical and economic merits of any mitigation strategy must be applied when selecting mitigation solutions. There is an extensive amount of information concerning mitigation techniques for harmonic distortion and indeed it would be possible to write an entire textbook on the highly complex field of harmonic filters. For this reason, this section provides an overview of mitigation options for harmonic distortion but does not delve into the complex aspects.

4.7.1 Network augmentation/changes to supply arrangements

As discussed previously, harmonic voltage distortion is due to the interaction between harmonic current and network impedance. As such, one method of reducing the magnitude of harmonic voltage is to reduce the network impedance. There are a number of practical methods, albeit in many cases expensive, that can be used to reduce network impedance and these include:

- Changing conductor to one of lower impedance. This will reduce the harmonic voltage distortion for any given harmonic current emission.
- Reorganising the supply arrangement such that the load is either supplied by a dedicated feeder or supplied from a higher voltage. This adjustment can reduce the impedance of the network and as such reduce the magnitude of voltage distortion for the same magnitude of harmonic current. Another advantage of changing the supply arrangement can be that it isolates the distorting load from other connections.

4.7.2 Harmonic filters

The purpose of harmonic filters is to either prevent the harmonic current from flowing through the network impedance, and as such reducing or eliminating the harmonic voltage or to eliminate the harmonic current entirely. Harmonic filtering is an option for both consumers and network operators. There are two filter technologies to be considered: passive filters and active filters.

In the case of passive filters, there are many designs in use (e.g. single tuned, double tuned, c-type). Regardless of the design, the theory behind passive filters is to provide a shunt circuit which presents as a low impedance to the harmonic orders which are to be filtered. In this case, the harmonic current will flow through this low impedance path instead of the higher network impedance path presented by the network resulting in lower voltage distortion. Passive harmonic filtering is achieved by designing a combination of capacitors and inductors that exhibit a series resonance (i.e. very low impedance) at the harmonic order which is to be filtered. In some cases it may also be possible to tune power factor correction capacitor banks to provide harmonic filtering. The process for this is similar to that used to detune the capacitor back as described in Section 4.4.1.4.

Active harmonic filters are power electronic devices which utilise switching components in order to generate waveforms which are completely out of phase for the harmonic order to be filtered. These out-of-phase waveforms will cancel the harmonic current thus eliminating the distortion. Active harmonic filters are more flexible than passive filters and also have the advantage of not altering the network impedance. However, they can be difficult to implement and are more expensive than passive filters.

4.7.3 Series reactors

When considering high emission sources such as variable speed drives, inclusion of a series reactor (inductor) can be a highly effective method of reducing waveform distortion. The series inductor tends to 'smooth' the waveform. A rule of thumb for sizing the series reactor is that it should be rated at 3% of the rating of the load.

4.7.4 Transformer phase shift

Another common method of reducing the harmonic emissions from industrial plant is the use of transformers with multiple windings designed to produce phase shifts in the secondary current. Through use of these devices, it is possible to reduce the emissions at various harmonic orders. Devices connected in this fashion are often called multipulse converters. For example, the current emission for a standard three-phase diode bridge rectifier circuit will have characteristic harmonic currents at orders $6n \pm 1$ (i.e. 5, 7, 11, 13, etc.). A phase shifting transformer arrangement can be designed to eliminate the 5th and 7th order emissions. This arrangement is termed a 12-pulse converter and the characteristic harmonics are of order $12n \pm 1$ (i.e. 11, 13, 23, 25, etc.). It is possible to have 24- and even 48-pulse converters which will effectively eliminate all low order harmonic emissions.

References

[1] IEC, Electromagnetic compatibility (EMC) part 4-7: testing and measurement techniques—general guide on harmonics and Interharmonics measurements and instrumentation for power supply systems and equipment connected thereto, IEC 61000-4-7:2002, 2002.
[2] S. Elphick, V. Smith, V. Gosbell, S. Perera, Characteristics of power quality disturbances in Australia: voltage harmonics, Aust. J. Electr. Electron. Eng. 10 (4) (2013) 490–496. 2013/01/01 2013.
[3] S. Elphick, V. Gosbell, V. Smith, R. Barr, Characteristics of power quality disturbance levels in Australia, in: IEEE 15th International Conference on Harmonics and Quality of Power (ICHQP) 2012 Hong Kong, 17–20 June 2012, 2012.
[4] J.F.G. Cobben, Power Quality—Problems and Solutions, CO-ED, Arnhem, The Netherlands, 2012.
[5] B.W. Kennedy, Power Quality Primer, McGraw Hill, USA, 2000.
[6] J. Gutierrez Iglesias, G. Bartak, N. Baumier, B. Defait, M. Dussart, F. Farrell, C. Graser, J. Sinclair, D. Start, M. Mazzoni, Power Quality in European Electricity Supply Networks—1st edition, Report Prepared for Eurelectric, 2002.
[7] A. Baggini (Ed.), Handbook of Power Quality, John Wiley & Sons, West Sussex, England, 2008.
[8] R. Barr, V.J. Gosbell, S. Perera, The customer benefits of high reliability and high power quality, in: EESA Electricity 2005 Conference, Brighton-le-Sands, NSW, Australia, November, 2005.
[9] S. Papathanassiou, N. Kasmas, N. Drossos, D. Stavropoulos, A practical evaluation of distribution network losses due to harmonics, in: 19th International Conference on Electricity Distribution (CIRED 2007), Vienna, Austria, June, 2007, 2007.
[10] P. Caramia, G. Carpinelli, P. Verde, Power Quality Indices in Liberalized Markets, John Wiley & Sons, West Sussex, United Kingdom, 2009.
[11] A. Eberhard (Ed.), Power Quality, InTech, Rijeka, Croatia, 2011.
[12] W.T.J. Hulshort, J.F. Groeman, Energy Saving in Industrial Distribution Transformers, Report Prepared for Kema, reference 40130017-TDC 02-24226A, 2002.
[13] K. Najdenkoski, G. Rafajlovski, V. Dimcev, Thermal aging of distribution transformers according to IEEE and IEC standards, in: Power Engineering Society General Meeting, 2007. IEEE, 24–28 June 2007, 2007, pp. 1–5.
[14] E.F. Fuchs, D.J. Roesler, K.P. Kovacs, Aging of electrical appliances due to harmonics of the power system's voltage, IEEE Trans. Power Delivery 1 (3) (1986) 301–307.

[15] IEEE, IEEE Standard for Harmonic Control in Electric Power Systems, IEEE Std 519–2022 (Revision of IEEE Std 519–2014), 2022, pp. 1–31.
[16] LG LRFC25750ST Service Manual, Webpage. Available from: http://www.manualslib.com/manual/685199/Lg-Lrfc25750st.html (Last Accessed 16 October 2022).
[17] IEC, Electromagnetic Compatibility (EMC)—Part 2-2: Environment—Compatibility Levels for Low-Frequency Conducted Disturbances and Signalling in Public Low-Voltage Power Supply Systems, IEC 61000-2-2:2002, 2002.
[18] Standards Australia, Electromagnetic Compatibility (EMC)-Limits—Assessment of Emission Limits for the Connection of Distorting Installations to MV, HV and EHV Power Systems, TR IEC 61000-3-6:2012, 2012.
[19] FinGrid, Power Quality in Fingrid's 110 kV grid, 2015.
[20] Operador Nacional do Sistema Eléctrico, Instruções para realização de estudos e medições de qee relacionados aos acessos à rede básica ou nos barramentos de fronteira com a rede básica para parques eólicos, solares, consumidores livres e distribuidoras, 2018.
[21] IEEE, IEEE Recommended Practice and Requirements for Harmonic Control in Electric Power Systems, 2014.
[22] Hydro-Québec, Emission Limits for the Disturbances on the Hydro-Québec Transmission System, 2019.
[23] RTE, Documentation technique de référence—article 8.3 Cahier des charges des capacités constructives pour une installation de production raccordée au RPT, 2017.
[24] D-A-CH-CZ, Technical Rules for the Assessment of Network Disturbances, 2007.
[25] T. Browne, V. Gosbell, R. Barr, Critical review of harmonic assessment procedures for transmission customers and renewable generators, in: CIGRE 2022 Paris Session, Paris, France, 2022.
[26] A. Božiček, J. Kilter, T. Sarnet, I. Papič, B. Blažič, Harmonic emissions of power-electronic devices in different transmission-network operating conditions, IEEE Trans. Ind. Appl. 54 (5) (2018) 5216–5226.
[27] A. Božiček, B. Blažič, E. Bećirović, J. Milanović, M. Löper, J. Kilter, T. Trummal, S. Hellmuth, M. Kuschke, K. Strunz, Influence of PQ disturbances on operation of PE rich power networks, Report Prepared for MIGRATE 31/12/2018, 2018.
[28] CIGRE—JWG C4/B4.38, Network modelling for harmonic studies, Report Prepared for April 2019, 2019.
[29] International Electrotechnical Commission, Wind Energy Generation Systems—Part 21-3: Measurement and Assessment of Electrical Characteristics—Wind Turbine Harmonic Model and Its Application, 2019.
[30] M. Pourarab, J. Meyer, M. Halpin, Z. Iqbal, S. Djokic, Interpretation of harmonic contribution indices with respect to calculated emission limits, in: 19th International Conference on Harmonics and Quality of Power (ICHQP), Dubai, United Arab Emirates, 2020.
[31] M. Halpin, Conceptual changes to the IEC TR series 61000-3-6, -7, -13 and -14 for power quality disturbance limit allocation procedures, in: IEEE International Conference on Industrial Electronics for Sustainable Energy Systems (IESES), Hamilton, New Zealand, 31 Jan - 02 Feb 2018, 2018.
[32] T. Tenne, Harmonic emission limits—position paper V2, Report Prepared for 15/10/15, 2015.
[33] Energy Networks Association, Planning Levels for Harmonic Voltage Distortion and the Connection of Non-Linear Equipment to Transmission Systems and Distribution Networks in the United Kingdom, 2019.
[34] EirGrid, Power Quality Requirements for Connection to the Transmission System, 2015.

[35] IEC, Electromagnetic Compatibility (EMC)-Limits—Assessment of Emission Limits for the Connection of Distorting Installations to MV, HV and EHV Power Systems, 2012.
[36] V.J. Gosbell, D.A. Robinson, S. Perera, A. Baitch, The application of IEC 61000-3-6 to MV systems in Australia, Quality and Security of Electrical Supply Conference Proceedings, ERA Technology, UK, 2001, pp. 7.1–7.1.10.
[37] V. Gosbell, Harmonic allocation to MV customers in rural distribution systems, in: Australasian Universities Power Engineering Conference (AUPEC) Perth, WA, Australia, 9–12 December 2007, 2007.
[38] V. Gosbell, D. Robinson, Allocating harmonic emission to MV customers in long feeder systems, in: Australasian Power Engineering Conference, Christchurch, October 2003, 2003.
[39] Standards Australia, HB-264, Power Quality—Recommendations for the Application of AS/NZS 61000.3.6 and AS/NZS 61000.3.7, 2003.
[40] V.J. Gosbell, R.A. Barr, A new approach to harmonic allocation for medium-voltage installations, Aust. J. Electr. Electron. Eng. 10 (2) (2013) 149–156.
[41] S. Elphick, V. Gosbell, V. Smith, R. Barr, The Australian long term power quality survey project update, in: 14th International Conference on Harmonics and Quality of Power, ICHQP'10, Bergamo, Italy, 26–29 September 2010, 2010.
[42] Australian Energy Market Operator, Integrated System Plan, Report Prepared for July 2020, 2020.
[43] Energy Networks Association (ENA), Guideline for Power Quality: Harmonics. Recommendations for the Application of the Joint Australian/New Zealand Technical Report TR IEC 61000.3.6:2012, 2014.
[44] E.F. Fuchs, D.J. Roesler, M.A.S. Masoum, Are harmonic recommendations according to IEEE and IEC too restrictive? IEEE Trans. Power Delivery 19 (4) (2004) 1775–1786. October, 2004.
[45] V. Ćuk, J.F.G. Cobben, P.F. Ribeiro, W.L. Kling, A review of international limits for harmonic voltages and currents in public networks, in: 16th International Conference on Harmonics and Quality of Power (ICHQP), Bucharest, Romania, 25-28 May 2014, 2014.
[46] M. Domagk, J. Meyer, M. Hoven, K. Malekian, F. Safargholi, K. Kuech, Probabilistic comparison of methods for calculating harmonic current emission limits, in: IEEE Manchester PowerTech, Manchester, United Kingdom, 18-22 June 2017, 2017.
[47] K. Malekian, F. Safargholi, W. Schufft, T. Dreyer, J. Cassoli, S. Adloff, F. Ackermann, H. Moghadam, S. Rogalla, G. Kaatz, M.F. Meyer, S. Mueller, I. Athamna, S. Daniel, R. Klosse, F. Santjer, B. Weise, Harmonic model validation of power generation units, IET Renew. Power Gener. 14 (13) (2020) 2456–2467.
[48] A.E. Emanuel, On the assessment of harmonic pollution [of power systems], IEEE Trans. Power Delivery 10 (3) (1995) 1693–1698.
[49] B. Peterson, J. Rens, G. Botha, J. Meyer, J. Desmet, Evaluation of harmonic distortion from multiple renewable sources at a distribution substation, in: IEEE International Workshop on Applied Measurements for Power Systems (AMPS), Liverpool, UK, 20–22 September 2017, 2017.
[50] X. Wilsun, Y. Liu, A method for determining customer and utility harmonic contributions at the point of common coupling, IEEE Trans. Power Delivery 15 (2) (2000) 804–811.
[51] T. Pfajfar, B. Blažič, I. Papič, Harmonic contributions evaluation with harmonic current vector method, IEEE Trans. Power Delivery 23 (1) (2008) 425–433.

CHAPTER 5

Impact and management of voltage fluctuations, flicker and rapid voltage changes

5.1 Introduction

Power system voltage waveform and frequency constantly undergo variations due to a range of reasons including connection and disconnection of loads and generators, faults, normal operation of loads such as electric arc furnaces, electric traction systems, welders and transformer energisation. The power system frequency variations are tightly controlled by the system operators whereas the voltage waveform at different locations in a power system can exhibit variations. With regard to fluctuations of the RMS voltage, they can be classified using the magnitudes, different timescales and averaging windows used to calculate them. In normal power system operation, if one considers the 10-min RMS value, the variations in the voltage waveform which happen in a continuous basis will be smoothed out whereas if half-cycle RMS values are considered, it is often possible to see the voltage variations as shown in Fig. 5.1. The preceding 10-min value can be related to steady-state voltage where the power system operators attempt to maintain it within limits, e.g. $\pm$ 10%, through the application of various procedures and techniques. The voltage fluctuations of concern in power systems from a quality of supply perspective considered in this chapter are associated with the modulating effect of the steady-state voltage that can be derived by employing the half-cycle RMS values as shown in Fig. 5.1.

Although rare, voltage fluctuations are said to be repetitive and rectangular if the voltage changes exhibit a behaviour, as shown in Fig. 5.2, indicating a cyclic modulating effect on the steady-state voltage. In this case, half-cycle RMS voltages remain the same over a cycle of the power system frequency and hence the fundamental frequency of modulation is half that of the power system frequency. In other words, there are two voltage changes in every cycle of the fundamental modulating frequency.

Applied Power Quality
https://doi.org/10.1016/B978-0-323-85467-2.00007-X

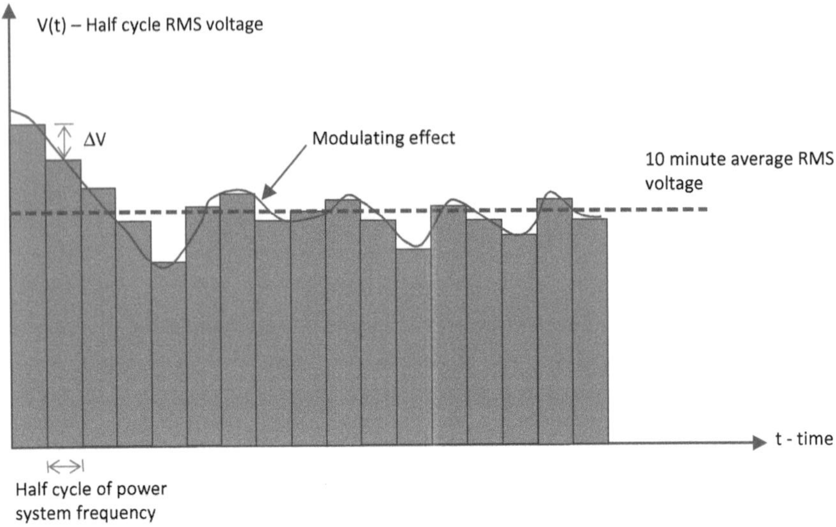

Fig. 5.1 Half-cycle RMS voltage variation with time and the modulating effect of the steady-state voltage.

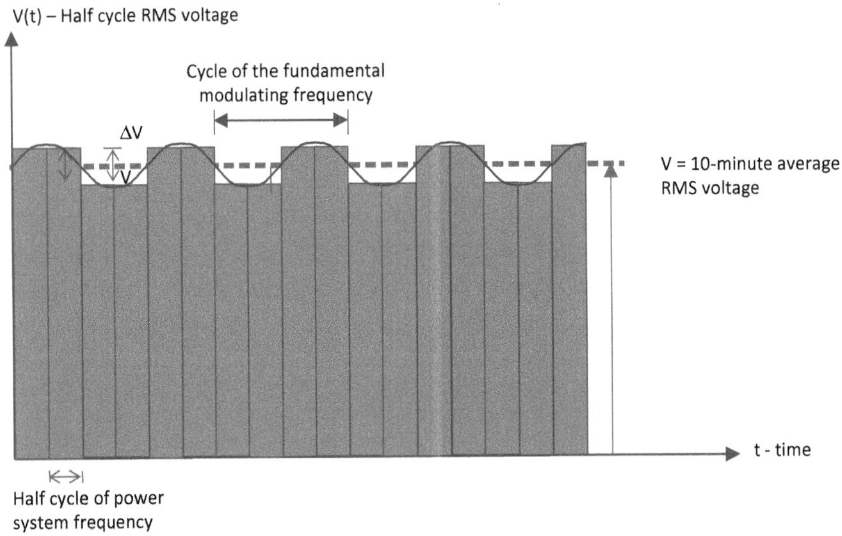

Fig. 5.2 Rectangular and repetitive variation of RMS voltage with time.

Even if voltage variations do not occur in a continuous manner as in Figs 5.1 and 5.2, an event such as a transformer energisation or a direct on-line motor start can lead to a change in the RMS voltage from one level to another level over a few cycles of the power system frequency, which is also classified as a voltage fluctuation.

5.2 Effects of voltage fluctuations

Under the normal operating conditions of power systems, the modulated voltage envelopes as shown in Figs 5.1 and 5.2 resulting from the operation of the various installations have little or no impact on the operation of many of the connected equipment, as the modulation amplitudes are relatively very small compared to the steady-state voltage. In addition, the associated modulating frequencies are also of little or no concern. However, the main impact of the modulated voltages is that they can lead to variations in luminance output of lamps—known as light flicker. Depending on the lamp technology, the perception of flicker is governed by the amplitude of the voltage envelope and its frequency and the individuals who perceive the flicker. The phenomena of light flicker and physiology behind flicker perception are well covered in the literature [1]. Referring to Fig. 5.2, the human eye may perceive light flicker associated with voltage fluctuations as small as 0.3% (i.e. $\Delta V/V$) if they occur at a frequency around 8 Hz which is equivalent to approximately 960 voltage changes/minute noting that there are two such changes in a voltage modulating cycle. In general, flicker becomes quite problematic if it is continuous as it can lead to irritation of the eye and sometimes can cause headache. Such continuous voltage variations are commonly associated with loads such as electric arc furnaces as the steel scrap melting phase can last over 30 min.

Traditionally light flicker has been associated with incandescent lamps which in fact have been phased out in many countries (although many remain in service). Lighting technologies have undergone significant changes over the years with more energy saving types such as compact fluorescent lamps (CFLs) and more recently light emitting diode (LED) lamps. These lamp technologies in general are less sensitive to voltage fluctuations as there is some energy storage in the DC link which forms an important part of the electronic circuitry which drive the lamps. However, sensitivity is dependent on the design of the electronic driving circuitry [2,3]. LED lamps are often used with dimmers and in such situations the combination can become sensitive to voltage fluctuations where the sensitivity tends to increase with the increasing dimming level—again depending on the design of the dimmer and the driving circuitry of the LED lamps [3].

Whilst continuous light flicker is the main concern associated with voltage fluctuations caused by loads such as electric arc furnaces and welders, even infrequent changes in light intensity from lamps may be of concern depending on the environment [4]. In rare occasions, some voltage fluctuations may lead to issues such as interference with communication equipment, spurious tripping of relays leading to equipment and process shut down, and interference with electronic equipment and with control systems. Voltage fluctuations, if significant, can lead to increase in losses in mains-connected induction motors, increase in ripple current in electronics, thus leading to temperature rise and degradation of life. Large voltage fluctuations can also cause damage to equipment.

5.3 Power quality parameters associated with voltage fluctuations

The two key power quality issues which fall under the category of voltage fluctuations are flicker and rapid voltage changes (RVCs). IEC61000-4-30, Ed 3:2015 [5] which covers testing and measurement methods of power quality parameters refers to IEC61000-4-15 [6] with regard to flicker measurement. The measurement of RVCs is clarified in IEC6100-4-30, Ed 3:2021 Amendment 1 [5] and replaces the content in IEC61000-4-30, Ed 3:2015.

5.3.1 Flicker measurement and indices

The measurement of flicker requires transformation of the voltage fluctuations incorporating the human physiological visual perception which involves the lamp–eye–brain combination. Considering traditional gas-filled, coiled tungsten lamps, the flicker perception range is considered to fall within the voltage modulating frequency range 0.05–35 Hz. The nature of the human eye is such that it is most sensitive to voltage fluctuations in the range 8–10 Hz whereas on either side of this range relatively larger voltage fluctuations are required for perception of flicker. Noting that over the entire flicker perception range of modulation frequencies, the measurement instrument should give the same output when flicker is perceived and hence frequency-dependent weighting factors (through a weighting filter) should be applied to the input signal which in fact is a voltage. This process and other steps and algorithms that should be followed are incorporated in the IEC flickermeter described in Ref. [6] and based on the 60-W gas-filled, coiled tungsten lamp. The modified weighting factors are given for lamps designed to operate at different voltages (230 V/50 Hz, 60 Hz, 120 V, 50 Hz, 60 Hz) noting that 120-V, 60-W lamps have thicker filaments and hence greater thermal inertia which means they require relatively larger voltage fluctuations to make them flicker. In this regard, the well-known flicker sensitivity curve, also commonly known as $P_{st} = 1$ flicker curve, refers to the threshold of

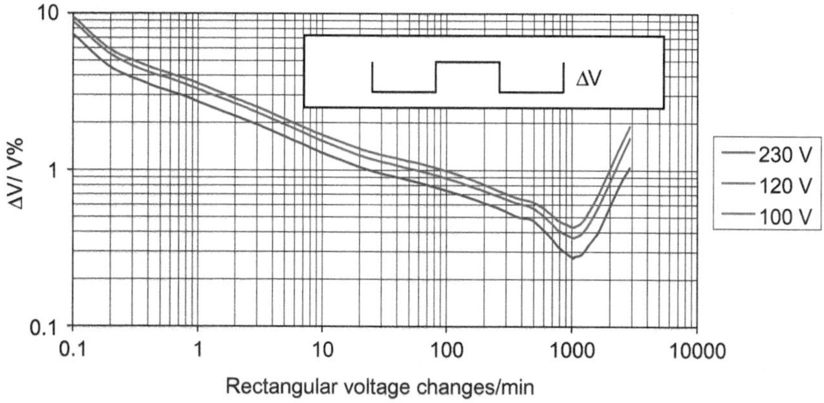

Fig. 5.3 Threshold of flicker perceptibility ($P_{st} = 1$ curve) with rectangular voltage fluctuations.

perception which illustrates the corresponding variation of relative voltage changes with voltage fluctuating rate considering rectangular voltage variations as shown in Fig. 5.3 [1] where 50% or more of the subjects will perceive lamp flicker associated with incandescent lamp considered.

Newer lamps such as LEDs possess voltage fluctuation sensitivity characteristics and hence flicker characteristics which are different when compared to traditional tungsten filament lamps. Where such modern lighting systems and lamps are prevalent, it has been suggested that the limiting value of $\Delta V/V$ in the voltage fluctuating frequency range be raised to a higher but a constant level in the frequency range 5–15 Hz [3].

Arising as a result of new lamp technologies, there is conjecture on the usefulness of the currently used IEC flickermeter algorithm that was developed based on the tungsten filament lamps. For this reason, there has been suggestions to make the flickermeter independent of the type of lamps used by making it light based [7] where the work is continuing. As stated previously, although LED lamps are considered to be less sensitive to voltage fluctuations, when they are operated with lamp dimmers of some designs, they can become more sensitive to voltage fluctuations compared to traditional gas-filled, coiled tungsten lamps. Until such time where newer flicker measurement techniques are developed and totally accepted through standardisation processes, the traditional IEC flickermeter will continue to be used. This is also because voltage fluctuations not only cause lamp flicker but are also considered to have an impact on other devices that are connected to the power system, as stated previously. Such devices that have been identified as experiencing operational and subsequent ageing issues include the electrolytic capacitors of power supplies, transformers, induction motors, adjustable speed drives, turbo alternators, phase-locked loops and power electronic converters [3] although there is no widespread evidence to believe that such issues are more dominant than lamp flicker. Hence, until consensus is reached internationally, any relaxation of $P_{st}=1$ limit that is based on the incandescent lamp is seen to be premature.

The IEC flickermeter provides the two quantities P_{st} and P_{lt} which are known as the short- and long-term flicker severity indices, respectively (which characterise flicker). P_{st} is measured over a 10-min window whereas P_{lt} is derived using 12 consecutive P_{st} indices as given by Eq. (5.1):

$$P_{lt} = \sqrt[3]{\frac{1}{12}\sum_{i=1}^{12} P_{sti}^3} \tag{5.1}$$

where P_{sti} ($i=1, 2,\ldots, 12$) are the 12 consecutive values of P_{st}.

Naturally, P_{lt} is a smoothed quantity and has been developed as flicker perceived over a longer period (i.e. 2 h) which could be a cause of concern compared to that observed over a 10-min window. Fig. 5.4 illustrates an example case of P_{st} and P_{lt} values recorded over a 7-day period (including the furnace down day) at a dedicated 132-kV busbar of a 330-kV/132-kV substation which supplies an electric arc furnace, the supply arrangement of which is shown in Fig. 5.5 [8]. As evident from Fig. 5.4, it is vital to note that a single high P_{st} value can dominate its effect on the P_{lt} over a period and it is important to investigate the reason for the occurrence of such a high P_{st} as it could be associated with a network event instead of voltage fluctuations caused by the installation being monitored, in this case the electric arc furnace. Such abnormal P_{st} values require filtering to eliminate their impact on P_{lt} values.

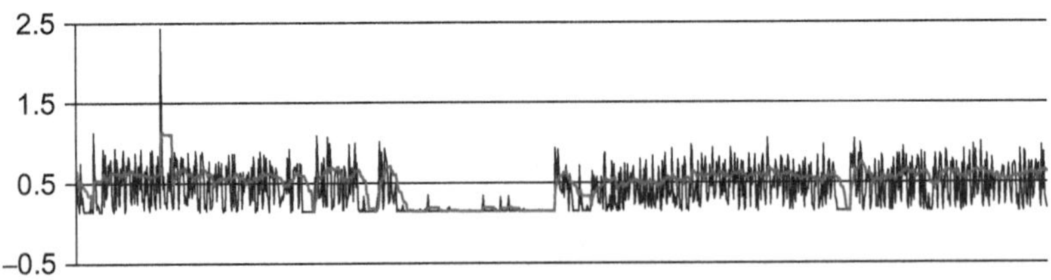

Fig. 5.4 Seven-day P_{st} and P_{lt} values recorded at the 132-V bus (of Fig. 5.5) which supplies an electric arc furnace.

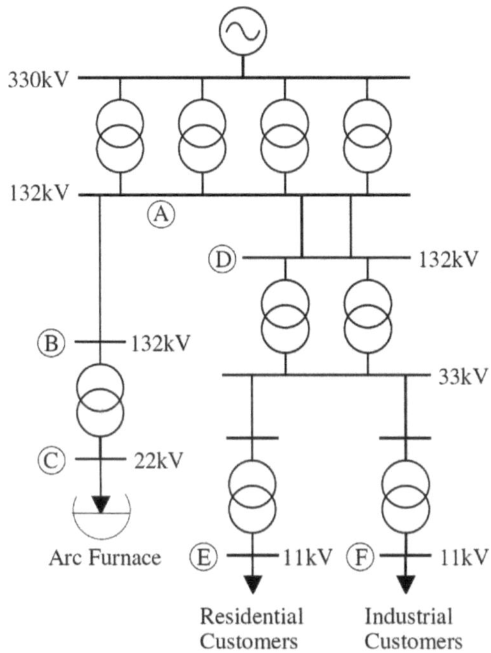

Fig. 5.5 Supply arrangement for an electric arc furnace installation supplied via a dedicated line [8].

5.3.2 Measurement of rapid voltage changes

A rapid voltage change (RVC) is broadly described as an abrupt change of RMS voltage from one steady state to another steady state. However, the recognition of an RVC and its measurement require attention to the detailed process described in IEC6100-4-30, Ed 3:2021 Amendment 1 [5]. The abruptness of the above change plays a key role in identifying an RVC. For illustrative purposes, three example RVC events are shown in Fig. 5.6 [4,9] where the two steady-state levels are identified by the voltage levels V_{ss1} and V_{ss2}. For a voltage change to be classified as an RVC, the voltage change should be within the voltage sag (dip) and swell thresholds.

An RVC event is characterised by the following four parameters as detailed in Refs. [5,9]:

- Start time
- Duration
- Maximum voltage change
- Steady-state voltage change

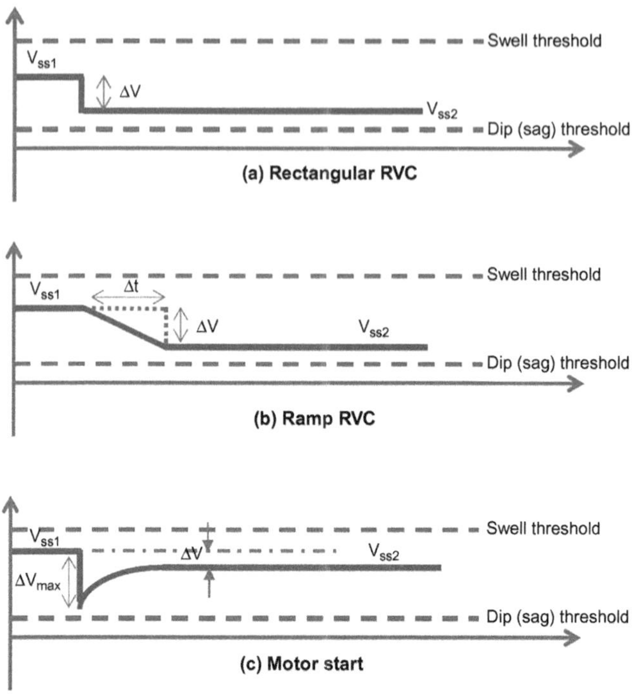

Fig. 5.6 Different types of rapid voltage changes [4,9].

For practical purposes and from the perspective of impact on equipment and processes, the maximum voltage change (ΔV_{max}) may be the quantity of interest (as in the case of the motor start) rather than the magnitude of the steady-state voltage change ($|V_{ss1} - V_{ss2}|$).

5.3.3 Flicker propagation and attenuation in power systems

Similar to other power quality disturbances such as harmonics and voltage unbalance, voltage fluctuations and flicker also tend to propagate from their sources to other locations in power systems. In this process, voltage fluctuations and flicker tend to often attenuate in their magnitude. Extensive practical and theoretical details on such propagation and attenuation aspects are covered in the literature [8,10,11]. These aspects become relevant in the management of voltage fluctuations and flicker in power systems including in their emission allocation as per the stipulated guidelines such as those given in IEC61000-3-7:2008 [12].

With regard to the propagation of voltage fluctuations caused by an installation, the well-known approximate expression (5.2) which gives the per unit voltage change ΔV at a connection point of an installation can be used, where the change in complex power of the installation is $(\Delta P + j\Delta Q)$ PU and its connection point is characterised by the per unit impedance $(R+jX)$.

$$\Delta V \approx \Delta P \cdot R + \Delta Q \cdot X \tag{5.2}$$

To illustrate the application of Eq. (5.2) consider the case where the impedance at the receiving end of a feeder is $(0.4+j0.8)$ per unit on a 100-MVA base where an induction motor is to be connected to the end of the feeder and started direct on-line (DOL). The DOL rating of the motor is 3.3 MVA at 0.3 pf lagging. The percentage voltage change at the receiving end of the feeder at motor start is as follows:

$$\Delta V \approx \left(\frac{3.3}{100} 0.3\right) 0.4 + \left(\frac{3.3}{100} \cdot \sin\left(\operatorname{acos}(0.3)\right)\right) 0.8 = 2.9\%$$

This voltage change will propagate towards the supply source along the feeder but with some attenuation, and for practical purposes the above voltage change can be related to fault levels along the feeder.

Further, for a feeder with relatively small R compared to X, Eq. (5.2) can be simplified as follows:

$$\Delta V \approx \frac{\Delta Q}{\text{FL}} \tag{5.3}$$

where FL represents the per unit fault level at the connection point, simply demonstrating that the voltage fluctuations can be minimised by connecting the installation at a location where the fault level is relatively large or a strong point in the network.

Installations such as electric arc furnaces are generally supplied by dedicated feeders (often radial) supplied by high voltage busbars of which the fault levels are relatively large whereas at the connection point of such installations, voltage fluctuations and hence the flicker will be substantially large as the fault levels are relatively small compared to the upstream high voltage busbars. This can be illustrated using the practical network considered in Ref. [8] shown in Fig. 5.5 using the concept of downstream to upstream flicker transfer coefficient ($T_{PstCB} = \frac{P_{stB}}{P_{stC}}$) (i.e. flicker transfer from busbar C to busbar B). At the same time, this network also illustrates the upstream to downstream flicker transfer coefficients ($T_{PstAE} = \frac{P_{stE}}{P_{stA}}, T_{PstAF} = \frac{P_{stF}}{P_{stA}}$). Based on Eq. (5.3), in relation to the radial line that supplies the arc furnace site of Fig. 5.5, the voltage fluctuations and flicker caused by the furnace will significantly attenuate as they propagate from busbar C to A. Based on the respective fault levels of the busbars C and A (400 and 7000 MVA, respectively) the calculated T_{PstCA} is 0.057, which has been noted to correlate well with the same established using scatter plots developed using synchronised flicker measurements [8]. From a flicker compliance perspective, the electric arc furnace, busbar A is considered to be the point of common coupling, where the level of attenuation is possible because of the transformer impedances (132/22 kV) and the 132-kV dedicated line impedance. However, any flicker levels that prevail on busbar A will propagate via the connected feeders to other downstream busbars, e.g. busbars D, E and F which serve other customers. It is known that such propagation can often happen with some attenuation where it will play a role with regard to the connection of other installations. If the level of attenuation is poor, then there will be less room for the connection of other disturbing installations. Theoretical [10,13,14], experimental [15] and field studies [8] demonstrate that the composition of the loads connected to the busbars plays a key role in flicker attenuation. The role of mains-connected three-phase induction motors has been specifically noted in this regard where under fluctuating supply voltage conditions (in the flicker frequency range), their effective impedance is somewhat smaller compared to the impedance with no fluctuations in the voltage. Although the associated process is complex, a rudimentary explanation of the voltage fluctuation attenuation due to such dynamic loads can be illustrated using Fig. 5.7, where V_{src} is the source voltage and V_{load} is the load voltage under normal operating conditions (with no voltage fluctuations) and Z_{line} and Z_{load} are the corresponding line and load impedances.

Assuming the fluctuating voltage components of the source and load voltages to be given by ΔV_{src} and ΔV_{load}, respectively, and the associated effective load impedance to be given by Z'_{load}, Eq. (5.4) can be written employing the voltage divider rule,

$$\frac{\Delta V_{\text{load}}/V_{\text{load}}}{\Delta V_{src}/V_{src}} = \left| \frac{Z'_{\text{load}}}{Z'_{\text{load}} + Z_{\text{line}}} \frac{Z_{\text{load}} + Z_{\text{line}}}{Z_{\text{load}}} \right| \tag{5.4}$$

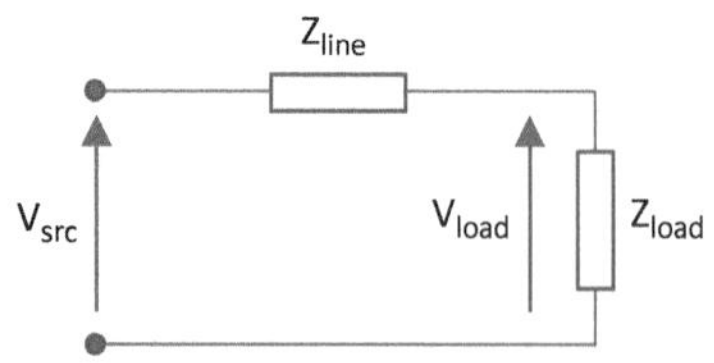

Fig. 5.7 Simple voltage divider rule which illustrates attenuation of voltage fluctuations.

The left-hand side of Eq. (5.4) can be considered to be proportional to the flicker transfer coefficient from the fluctuating source to the load. If static loads such as heating loads where $Z'_{load} \approx Z_{load}$ are considered, then the above transfer coefficient will be nearly equal to unity. However, under voltage fluctuating conditions the effective load impedance Z'_{load} will be smaller in magnitude compared to its static value such as in the case of three-phase induction motors; the above flicker transfer coefficient will be less than unity, thus explaining the associated flicker attenuation behaviour. At the same time if the loading level offered by the load is reduced (i.e. large Z_{load} and Z'_{load}) the flicker attenuation coefficient will approach unity.

In practical networks, usually a vast collection of mains-connected three-phase induction machines are present in both LV and MV levels which assist with flicker attenuation in a collective manner. This was also evident in the network of Fig. 5.5, where busbar F exhibited less flicker compared to busbar E where the former is classified as industrial (containing a large collection of three-phase induction motors) and the latter as residential (containing more passive loads than three-phase induction motors), noting that the flicker levels present at busbar A are common to both busbars E and F. This flicker attenuation behaviour and the corresponding flicker transfer coefficients were established using synchronised flicker measurements as detailed in Ref. [8].

This flicker attenuation behaviour is demonstrated through EMT simulations [16] undertaken using the 12.47-kV, 3.2187-km radial line considered in Chapter 3. The results of this simulation work are illustrated in Fig. 5.8. The sending end phase voltages ($V_{sending}$) of the line are sinusoidally modulated with a ΔV of approximately 0.36% at a modulating frequency (f_m) of 2 Hz. At the receiving end of the line, a group of 4-kV three-phase induction motors (with a total rating of 1.6 MVA each rated at 200 kVA) is connected through a step-down transformer. Their flicker attenuation behaviour is compared with a passive load with a complex power loading level (P_pas_tot and Q_pas_tot) equivalent to that of the motor load (P_mot_tot and Q_mot_tot). The 4-kV bus voltage ($V_{receiving}$) is observed initially with the passive load connected which is replaced with the motor load at 5.5 s.

From Fig. 5.8 (ignoring the transient behaviour at 5.5 s) it is evident that upstream voltage fluctuations do propagate to downstream busbar with some attenuation in both cases. However, when the downstream busbar is connected with induction motors, it demonstrates reduced voltage fluctuations compared to the case with passive load. This flicker attenuation

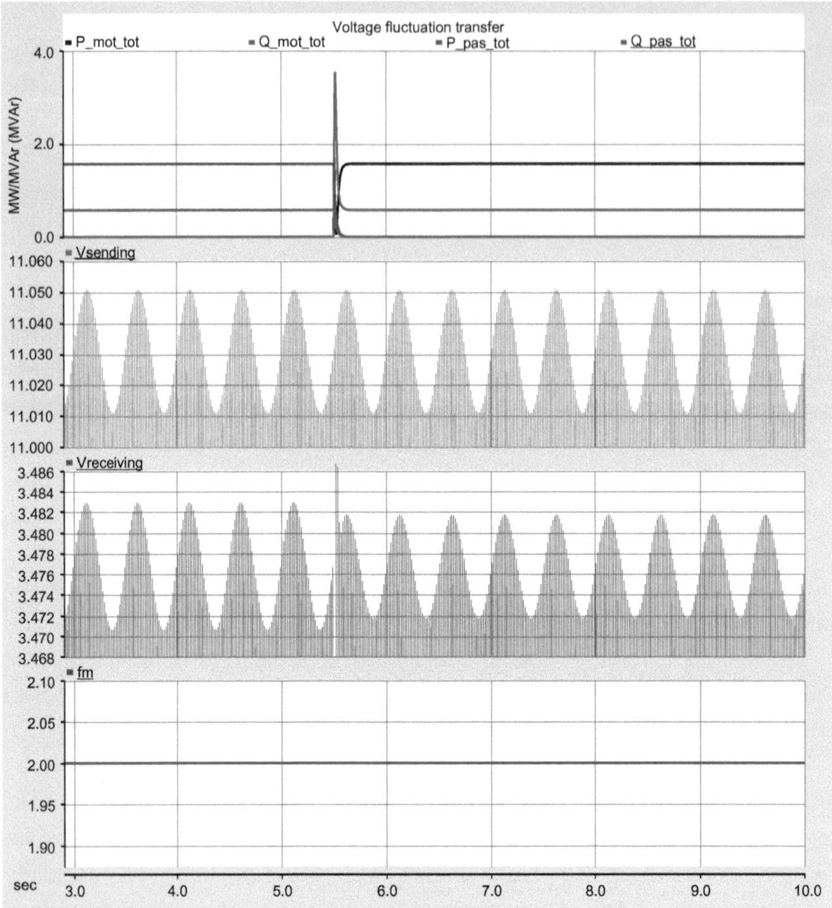

Fig. 5.8 Simulation results illustrating voltage (flicker) propagation and attenuation.

behaviour although has been extensively examined theoretically in Refs. [10,13,14], in practice, the fluctuating power system voltages (for example those arise as a result of electric arc furnaces) will consist of numerous frequency components and hence synchronised flicker measurements using flickermeters will provide an aggregate behaviour. Fig. 5.9 illustrates the variation of the flicker transfer coefficient measured in a laboratory setup with a 2.2-kW three-phase induction motor load and an equivalent passive load connected at the downstream location B of an emulated radial network of which the supply end A was sinusoidally modulated [15]. With the induction motor, whilst noticing flicker attenuation, the measured flicker transfer coefficient T_{PstAB} illustrates clearly its dependency on the modulating

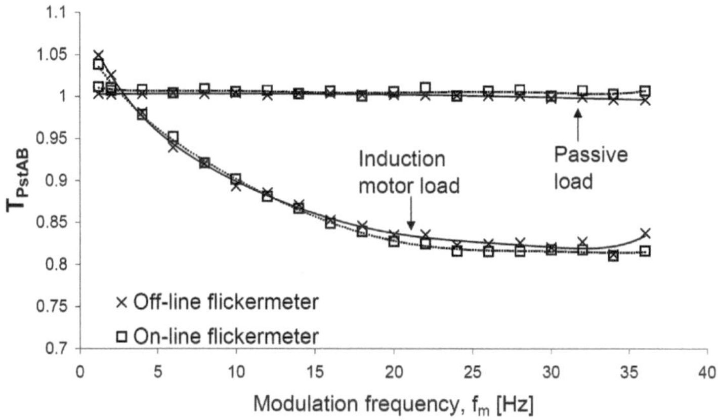

Fig. 5.9 Laboratory results demonstrating flicker attenuation behaviour associated with mains-connected induction motors [10].

frequency whereas the passive load does not contribute to any visible or significant attenuation. This experimental investigation further consolidates the role played by induction motors in comparison to passive loads in power systems.

5.4 Management of voltage fluctuations and flicker and their measurement and network planning levels

As stated previously, many power system fluctuating loads are supplied by dedicated lines (e.g. electric arc furnaces) or are connected to electrically strong points (i.e. with low impedance) and hence flicker is often not seen to be a significant problem. Based on vast amounts of data collected and analysed [11], it is evident that many of the measurements are made at HV/EHV levels whereas flicker is usually a problem in LV networks. The poor correlation between problematic flicker levels which exceed voltage level-specific planning levels and lack of complaints from LV network users can be attributed to significant flicker attenuation when it transfers from networks operating at higher voltages to lower voltage networks. The poor correlation is further aided by the significant changes in lighting technologies which have taken place over the years by moving away from traditional incandescent lamps to other types, although the flicker magnitudes may well exceed the compatibility level of $P_{st} = 1.0$ in low voltage networks.

With regard to problematic loads such as electric arc furnaces, empirical formulae are available to predict the resulting flicker that will arise at their design stages [1]. There are also approximate methods that could be applied for the calculation of flicker produced by loads such as welders which are characterised by periodic load cycles that lead to well-defined rectangular or other RMS voltage changes [1,12]. Such approximate approaches are useful noting that the algorithm of the IEC flickermeter follows a complex procedure making it difficult to apply in manual calculations.

As per the IEC philosophy, to ensure electromagnetic compatibility in networks, flicker compatibility limits are specified for low voltage networks [17] ($P_{st}=1.0$, $P_{lt}=0.8$) so that flicker planning levels can be adopted in MV, HV and EHV [12] and incorporated into the design of equipment. Such a set of planning levels are given in Table 5.1. With the availability of network-specific flicker transfer coefficients, it is possible to revise these planning levels whilst ensuring the compatibility levels at LV level [12].

The planning levels given in IEC61000-3-7:2008 [12] for rapid voltage changes are shown in Table 5.2. A summary of the limits applied in several countries is given in Ref. [9]. The limits associated with relatively infrequent events such as transformer energisation are not addressed through limits. Such quite infrequent events can also be imposed with specific limits on a case-by-case basis.

Using the planning levels adopted, the determined global emission levels can be apportioned to installations [12]. Where an installation is seen to exceed the allocated flicker emission levels, adequate measures need to be taken to curtail the flicker emission levels within the installation itself by using technologies such as Static Var Compensators (SVCs) or Static Compensators (STATCOMs) [1].

Table 5.1 Indicative flicker planning levels [12].

Voltage level	P_{st}	P_{lt}
MV	0.9	0.7
HV-EHV	0.8	0.6

Table 5.2 Indicative planning levels for rapid voltage changes [12].

	$\Delta V/V\%$	
Number of changes (n)	**MV**	**HV/EHV**
$n \leq 4$ per day	5–6	3–5
$n \leq 2$ per hour and > 4 per day	4	3
$2 < n \leq 10$ per hour	3	2.5

Naturally, it is important to note that rapid voltage changes and flicker measurements should be carried out using compliant instruments and appropriate measurement methods [5,6]. In assessing an installation for compliance, it is vital that disturbances that arise outside the installation are removed as illustrated in relation to flicker measurements shown in Fig. 5.4.

Unlike in the measurement of harmonics where adequate bandwidth should be possessed by the instrument transformers, flicker can be measured using station auxiliary transformers, which are often, but not always, very lightly loaded. In such measurements, it is important that flicker that is caused by connection or disconnection of station equipment is filtered out. This goes to state that flicker being a low-frequency phenomena, the requirements on instrument transformers are not as stringent in harmonic measurement.

The resultant flicker (e.g. P_{st}) that arises as a result of different installations has been noted to follow a general summation law of the type:

$$P_{st} = \sqrt[\alpha]{\sum P_{sti}^{\alpha}} \tag{5.5}$$

where

P_{st} = resultant short-term flicker due to all emitting sources.

P_{sti}= short-term flicker caused by the ith flicker emitting source.

α =summation exponent.

The summation exponent of $\alpha = 3$ is largely used in practice where the risk of coincident voltage changes is very small but other exponents that can be used under other conditions are described in Ref. [12].

References

[1] Baggini., Handbook of Power Quality, John Wiley & Sons, 2008.

[2] APQRC Tech Note 15, Impact of Power Quality on Lighting Technology. http://eisweb.adeis.uow.edu.au/apqrc/content/technotes/APQRC%20TN015-1604%20Impact%20of%20PQ%20on%20Lighting.pdf.

[3] CIGRE Technical Brochure 656, Review of LV and MV Compatibility levels for Voltage Fluctuations, 2016.

[4] K. Brekke, H. Seljeseth, O. Mogstad, Rapid voltage changes—definition and minimum requirements, in: CIRED 2009—The 20th International Conference and Exhibition on Electricity Distribution—Part 2, 2009, pp. 1–2, https://doi.org/10.1049/cp.2009.0984.

[5] IEC 61000-4-30:2015+AMD1:2021 CSV Consolidated version, Electromagnetic Compatibility (EMC)—Part 4-30: Testing and Measurement Techniques—Power Quality Measurement Methods, 2021.

[6] IEC 61000-4-15:2010, Electromagnetic Compatibility (EMC)—Part 4-15: Testing and Measurement Techniques—Flickermeter—Functional and Design Specifications, 2010.
[7] J. Drapela, J. Slezingr, Design and utilization of a light flickermeter, in: 2012 IEEE International Workshop on Applied Measurements for Power Systems (AMPS) Proceedings, 2012, pp. 1–6, https://doi.org/10.1109/AMPS.2012.6344007.
[8] S. Perera, et al., Synchronized flicker measurement for flicker transfer evaluation in power systems, IEEE Trans. Power Deliv. 21 (3) (2006) 1477–1482, https://doi.org/10.1109/TPWRD.2005.864057.
[9] APQRC Tech Note 18, Rapid Voltage Changes. http://eisweb.adeis.uow.edu.au/apqrc/content/technotes/APQRC%20TN018-2106%20Rapid%20Voltage%20Changes.pdf.
[10] S. Tennakoon, Flicker Propagation in Radial and Interconnected Power Systems (PhD thesis), School of Electrical, Computer and Telecommunications Engineering, University of Wollongong, 2008. http://ro.uow.edu.au/theses/96.
[11] CIGRE Technical Brochure 449, Review of Flicker Objectives for LV, MV and HV Systems, 2011.
[12] IEC TR 61000-3-7:2008, Electromagnetic Compatibility (EMC)—Part 3-7: Limits—Assessment of Emission Limits for the Connection of Fluctuating Installations to MV, HV and EHV Power Systems, 2008.
[13] S. Tennakoon, S. Perera, D. Robinson, Flicker attenuation—part I: response of three-phase induction motors to regular voltage fluctuations, IEEE Trans. Power Deliv. 23 (2) (2008) 1207–1214, https://doi.org/10.1109/TPWRD.2007.908788.
[14] S. Tennakoon, S. Perera, D. Robinson, Flicker attenuation—part II: transfer coefficients for regular voltage fluctuations in radial power systems with induction motor loads, IEEE Trans. Power Deliv. 23 (2) (2008) 1215–1221, https://doi.org/10.1109/TPWRD.2008.915826.
[15] S. Tennakoon, S. Perera, D. Robinson, Attenuation of flicker by induction motor loas: a laboratory investigation, in: Proc. 12th International Conference on Harmonics and Quality of Power (ICHQP 2006), Cascais, Portugal, 2006.
[16] PSCAD™, Version 5, Manitoba Hydro International.
[17] IEC 61000-2-2:2002, Electromagnetic Compatibility (EMC)—Part 2-2: Environment—Compatibility Levels for Low-Frequency Conducted Disturbances and Signalling in Public Low-Voltage Power Supply Systems, 2002.

CHAPTER 6

Impact and management of power system voltage sags

6.1 Introduction

This chapter examines the causes and impacts of voltage sags. A definition for voltage sags is provided as well as a summary of the causes of voltage sags. Data collected from field measurements is then provided in order to give the reader an understanding of the actual (empirical) behaviour of voltage sags in the field.

The chapter then provides an example of how to calculate the impact of voltage sags. This is followed by an evaluation of the impact of voltage sags in terms of equipment performance, industry immunity curves, network capability and finally economic impact.

Mitigation options from both a network and equipment/installation perspective are then detailed. The chapter concludes by providing insight into methods of assessing and reporting voltage sags.

6.2 Definition of voltage sags

A voltage sag, sometimes known as a voltage dip, is a short-term reduction in the rms voltage. The IEC electrotechnical vocabulary, IEC 60050 [1], defines a voltage sag as any.

> *sudden reduction of the voltage at a point in the electrical system, followed by voltage recovery after a short period of time, from half a cycle to a few seconds [1].*

Voltage sags are generally characterised by their duration and depth. Duration is the length of time for which the voltage remains below a threshold. The concept of depth is somewhat a misnomer as a sag is characterised by the retained voltage, that is, the voltage which persists during the sag, as opposed to the voltage decrease or 'lost' voltage. Whilst the IEC definition does not give a set of definitive durations or level of retained voltage that must be met for a disturbance to be classified as a voltage sag, IEEE Standard 1159 [2]

Applied Power Quality
https://doi.org/10.1016/B978-0-323-85467-2.00004-4

defines a voltage sag as a variation in the rms voltage of duration greater than ½ a cycle and less than 1 min with a retained voltage of between 10% and 90% of nominal. This reduction of voltage to between 10% and 90% of the nominal (or reference) voltage and duration between ½ a cycle and 1 min is the generally accepted definition of a voltage sag. In terms of duration, any disturbance that persists for less than ½ cycle is considered transient phenomena whilst voltage variations or disturbances of duration greater than 1 min are considered to be either sustained undervoltage or interruptions. In terms of depth, retained voltage above 90% is considered to be within the normal operating range whilst depth below 10% retained voltage is an interruption.

Further to the above, the IEEE categorises sags/interruptions according to their duration. These categories are presented in Table 6.1. The IEEE duration categories are to some extent correlated with cause.

As intimated previously, interruptions, sometimes referred to as outages or power failures, are distinct from sags as for these events the voltage level effectively reaches zero. Interruptions may also last from fractions of a second up to hours or even longer. The primary causes of interruptions are network faults and planned outages. Fig. 6.1 shows an example of a voltage waveform containing a voltage sag. As can be seen, the voltage decreases for a number of cycles before returning to the pre-sag voltage magnitude.

Table 6.1 IEEE sag/interruption duration categories.

Term	Duration	Likely cause
Instantaneous	0.5 cyc–0.5 s	Fault
Momentary	0.5 s–3 s	Motor DOL start
Temporary	3 s–1 min	Recloser operations

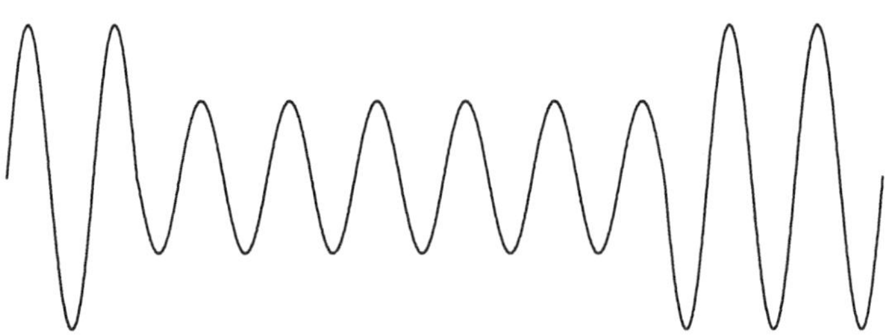

Fig. 6.1 Voltage sag waveform.

Voltage sags are measured by specialised power quality monitoring instrumentation. The instrument must be configured with a sag threshold voltage. That is, a voltage magnitude that will trigger a sag capture if/when the rms voltage falls below it. Fig. 6.2 shows a graphical representation of the rms voltage associated with a voltage sag, including the sag threshold and the parameters (duration, retained voltage) used to report/characterise the sag. Note the use of a hysteresis value in Fig. 6.2; this value is used to prevent voltage levels which are close to the sag threshold crossing the threshold multiple times and triggering multiple sags which are basically due to the same event.

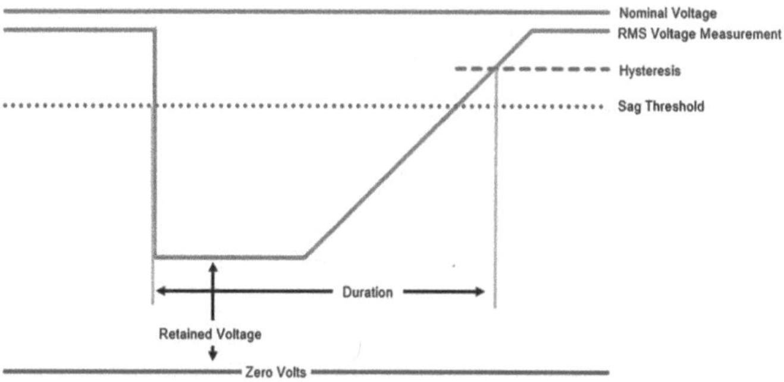

Fig. 6.2 Example of the RMS voltage trend for voltage sag, including key parameters—sag threshold, retained voltage and duration.

In practice, voltage sags are unlikely to have the relatively ideal characteristics shown in Figs 6.1 and 6.2. In real networks it is likely that the voltage will experience a phase angle jump due to the instantaneous change in system impedance. This phenomenon is illustrated in Fig. 6.3 where the dotted

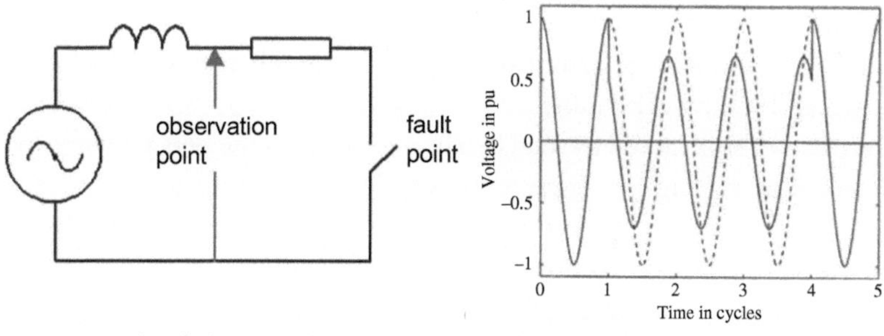

Fig. 6.3 Example of phase angle jump due to a voltage sag.

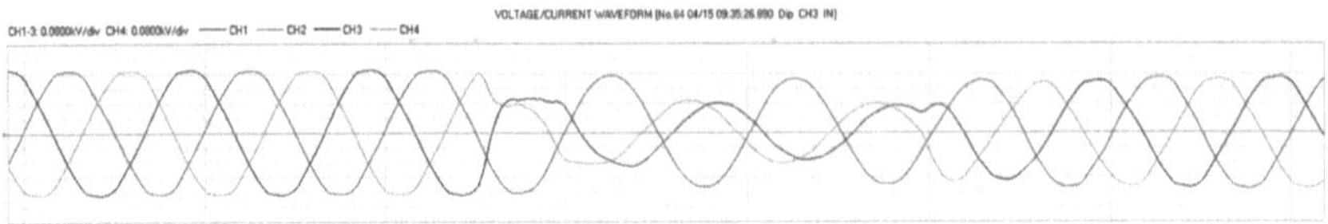

Fig. 6.4 Measured voltage sag waveform showing phase jump.

line on the waveform graphs shows the waveform without the sag and the solid line shows the sag voltage including the phase angle jump. In real circuits, the higher the network resistance, the higher will be the phase angle jump. Fig. 6.4 shows a measured voltage sag waveform including phase jump on one of the phases. Voltage sags due to faults may also 'evolve' as the fault develops, e.g. a single phase to ground fault may 'evolve' into a three-phase to ground fault. As such, the rectangular rms voltage profile displayed in Fig. 6.2 represents an idealised voltage sag.

6.3 Causes of voltage sags

The underlying cause of voltage sags is the interaction between large currents (i.e. multiple times the normal load current) and network impedances. The depth of the voltage sag is related to the magnitude of the current and the magnitude of the network impedance. The duration of the sag is related to the time for which the current persists. The two main causes of voltage sags are network faults and the energisation of equipment (loads) which draw large currents. In both cases, there are many factors which influence the sag envelope waveshapes, for example arc characteristic, earthing impedance, feeder R/X ratio, and motor and load characteristics.

For sags caused by network faults, the sag duration will be influenced by the time taken to clear the fault. This in turn is influenced by protection settings and characteristics. Recloser operation, which is used in an attempt to maintain supply in the event of a temporary fault, can also lead to additional sags. The nature of the fault, e.g. single line–ground, line–line, three-phase ground, will also have a significant impact on the characteristics of voltage sags caused by network faults.

In terms of sags caused by load behaviour, possibly the best known cause is direct-on-line starting of induction motors. During starting, these devices will draw multiple (up to 8) times the rated current. Energisation of transformers, particularly large power transformers, is another well-known source of voltage sags. Further detail concerning the behaviour of direct-on-line motor starting as well as transformer energisation is provided in Chapter 7.

6.4 Empirical characteristics of voltage sags

As discussed previously, the characteristics of voltage sags are influenced by a number of factors, including protection characteristics, load characteristics, type of fault and network characteristics. This section presents empirical findings which provide insight into the nature of voltage sags observed in practical low voltage distribution systems. This empirical insight has been derived through analysis of data collected as part of the Australian Long Term National Power Quality Survey (LTNPQS) project [3]. This data has been used to develop characteristics of voltage sags in Australia. The characteristics of the data which was analysed are as follows:

- Data was measured between 1st July 2004 and 30th June 2013.
- Data has been collected from 2556 distinct sites.
- The sites were predominately located at distribution (MV/LV) transformers and are connected to the LV terminals. The sites are supplied by a mix of network types (underground and overhead) and supply a mix of predominant loads, including residential, commercial and industrial.
- The data included for analysis was limited to voltage dips of residual voltage greater than 0% and less than 90% of nominal voltage (230 V). Duration of dips is restricted to being between 0.01 s (½ a cycle at 50 Hz) and 60 s (3000 cycles at 50 Hz).

Fig. 6.5 shows a histogram of the residual voltages obtained using the earlier data. The figure shows that the vast majority of voltage dips are quite

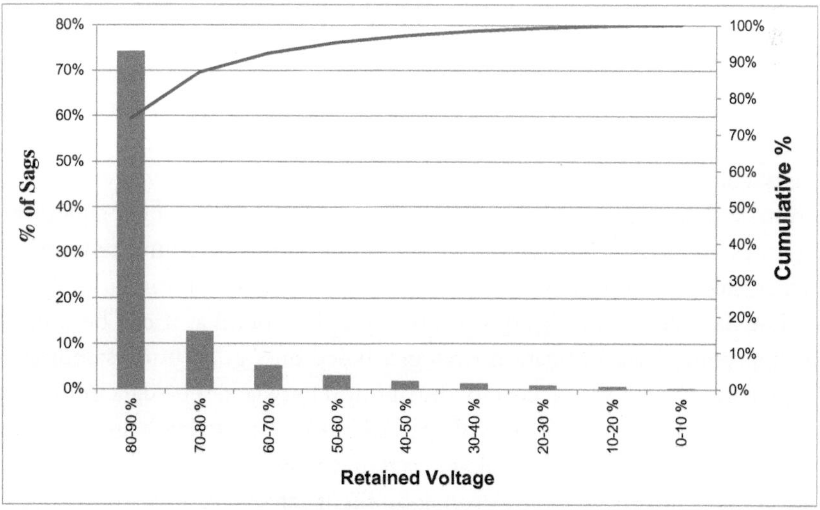

Fig. 6.5 Histogram of voltage sag retained voltage.

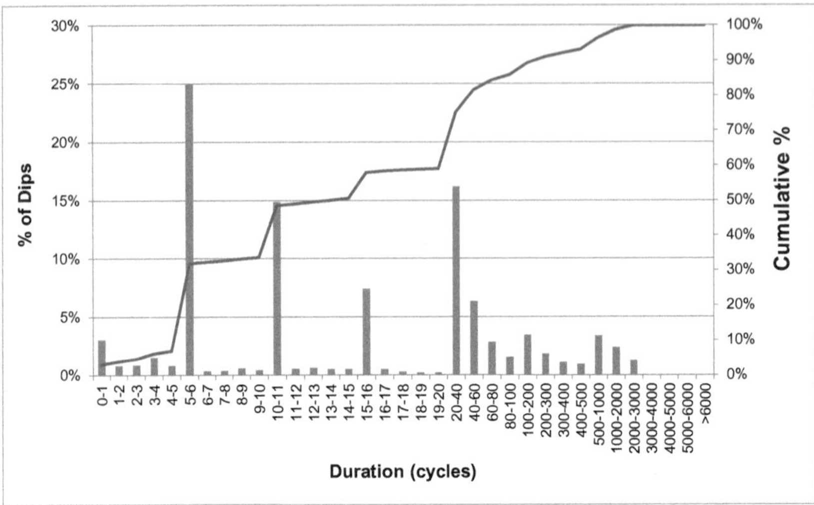

Fig. 6.6 Histogram of voltage sag duration.

shallow, being above 70% residual voltage. These dips are the result of either shallow reflected faults or possibly the starting of large loads, such as direct-on-line connected motors.

Fig. 6.6 shows a histogram of dip durations. Two distinct clusters of duration can be identified. These are the groups on the left of the graph and the groups on the right of the graph. The first cluster (group on the right) of short duration values (to approximately six cycles) has characteristics that indicate that they are faults in the upper MV or HV network (33 kV and above) which are cleared very quickly by unit protection and other fast acting transmission or subtransmission protection schemes. The second cluster (group on the left) of longer duration values has characteristics which can be attributed to one of two causes: voltage dips in other parts of the MV network which have slower over-current and distance protection schemes, and subsequently longer fault clearing times, or starting of large loads such as direct-on-line connected motors, such as those found in devices such as pumps and air conditioning units.

The data shown in Figs 6.5 and 6.6 is highly useful as it can be used to develop voltage sag mitigation strategies based on a good understanding of the performance of the electricity distribution network. The data in Fig. 6.5 shows that the vast majority of voltage sags have a retained voltage of greater than 80%. In fact, 82% of voltage sags have retained voltage of 50% or greater. This means that mitigation equipment capable of addressing voltage sags with depth down to 50% retained voltage will be effective in the vast majority of cases. Examination of sag durations indicates that 68% of voltage sags are of duration of 1 s or less. This figure gives an indication of the hold-up time required by mitigation devices if they are to be effective for most voltage sags. If the duration is extended to 2 s, 97% of voltage sags fall within this duration.

6.5 Factors influencing voltage sag severity

Proximity to power system faults plays a significant role in determining the depth of voltage sags and the duration is mostly dependent on the protection settings unless the faults are of the self-healing type. Power system faults are of varying types which include balanced or unbalanced. In practice, single-line-ground faults are the most common. Voltage sags associated with balanced faults are relatively easily to analyse whereas voltage sags associated with unbalanced faults are relatively difficult and analysis often requires the use of symmetrical components. With the availability of computer software, many of these tasks can be automated. However, an in-depth understanding of the details associated can be developed by carrying out hands-on calculations. The intention of this section is not to be exhaustive in relation to fault calculations (which are well covered in commonly available literature) but to illustrate the basic principles which can be used to determine the severity of voltage sags due to a fault.

These principles are illustrated by the following example:

Data: A 132 kV, bulk supply point in a power system has a three-phase fault level of 3000 MVA, and the associated impedance is characterised by an X/R ratio of 8:1. The supply feeds a 132 kV/11-kV, 25 MVA three-phase transformer with an impedance of 12% (on its own rating) which is connected to an 11 kV, 12 km overhead three-phase feeder having an equal resistance and reactance of 0.4 Ω/km.

Problem: For a balanced bolted three-phase fault (a) on the 11 kV secondary of the transformer (b) half way along the overhead feeder and (c) at the end of the overhead feeder calculate the sag voltage as a percentage of the nominal voltage ignoring prefault current in the feeder.

Solution

(a) For the fault on the secondary of the transformer, the voltage sag depth will be 100% and hence the sag voltage V_{sag} would be zero during the fault.

To answer Part (b) and (c) it is convenient to use the per unit method. Assuming a base MVA of 100 and 132 kV as the base values for the 132 kV side,

Per unit short circuit impedance Z_{HV} on the 132 kV is as follows:

$$|Z_{HV}| = \frac{100}{3000} = 0.333 \text{ pu}$$

Using the given X/R ratio, its complex representation is as follows:

$$Z_{HV} = (0.0413 + j0.3307) \text{ pu}$$

Per unit impedance of the transformer noting it is totally reactive:

$$X_{TF} = j0.12 * \frac{100}{25} = j0.48 \text{ pu}$$

The total feeder impedance in pu is as follows:

$$Z_{\text{feeder}} = (0.4 + j0.4) * 12 * \frac{100}{11^2} = (3.967 + j3.967)\text{pu}$$

Using the simple impedance divider rule:

(b) For the fault halfway along the feeder

$$V_{\text{sag}} = \left| \frac{(3.967 + j3.967)/2}{0.0412 + j0.3307 + j0.48 + (3.967 + j3.967)/2} \right| = 0.82 \text{ pu}$$
$$= 81\%$$

(c) For the fault at the end of the feeder

$$V_{\text{sag}} = \left| \frac{(3.967 + j3.967)}{0.0412 + j0.3307 + j0.48 + 3.967 + j3.967} \right| = 0.90 = 90\%$$

It is clearly evident that the further away that the fault is from the transformer secondary (or feeder front end) the higher is the retained voltage that will be observed by the customers on the other parallel feeders that are supplied by the transformer.

With unbalanced faults, the voltage sag (or swell) levels will depend on the star point grounding of the transformers. In the case of solidly grounded transformers, the healthy phases will maintain the prefault voltages and hence will not experience any voltage sags. Higher voltage networks (e.g. 132 kV) tend to be solidly grounded and hence depending on the location of short circuit on the faulted phase only that phase will experience the voltage sag where the depth of the sag will improve if the fault is far away from the transformer.

In the case of lower voltage networks (e.g. 11 kV) sometimes transformers are resistance or reactance grounded. In such cases whilst the faulted phase will experience the sag (sag depth depending on the location of the short circuit), the healthy phases will experience a voltage swell where the magnitude of the swell will depend on the location of the fault. The voltage swell will be greater if the fault happens closer to the transformer and it will be less severe if the fault occurs far away from the transformer secondary. Mathematical formulations associated with such situations are well covered in many books, e.g. [4].

6.6 Impact of voltage sags

6.6.1 Equipment susceptibility

As discussed previously, a voltage sag is characterised by its duration and voltage. These parameters can be represented by a point on sag duration/voltage plane. Using the same sag duration/voltage plane, equipment immunity can be illustrated as a boundary separating the points for which equipment will operate and points for which equipment will fail to operate as expected or trip entirely. This boundary is a method of showing the sag immunity characteristics of the equipment and is sometimes called a voltage tolerance characteristic. For many items of equipment, especially those with a diode/capacitor front ends, the immunity characteristic is often a rectangle similar to that shown in Fig. 6.7. The critical point shows the lowest voltage at which the device can operate indefinitely and the longest time for which it can operate at zero volts.

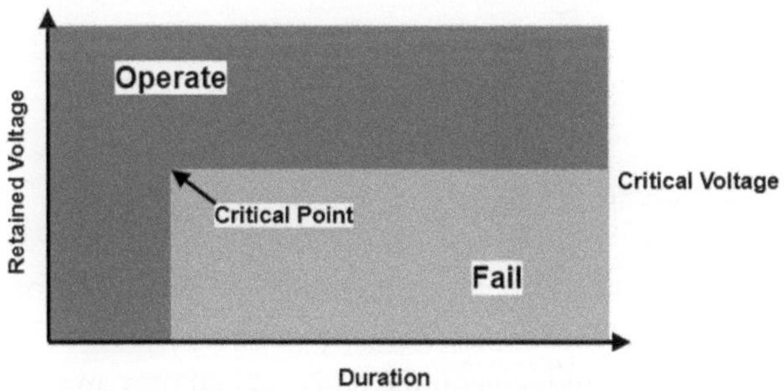

Fig. 6.7 Example of equipment sag immunity characteristic.

6.6.1.1 *Switch mode power supplies*

Switch mode power supplies (SMPS) are a common item at the front end of electronic equipment such as computers, TV and PLCs. With respect to sag performance, their important components are a full-wave bridge rectifier and a smoothing capacitor for the DC voltage applied to the next stage of the circuit. During a deep sag, the supply is unable to charge the capacitor whose voltage is then decreased by the current drawn by the next stage. If the sag duration is sufficiently long, the capacitor voltage will fall too low for proper operation and, depending on the design, the equipment will either trip or malfunction. The maximum sag duration allowing proper operation depends on the capacitor size, the current drawn and the range of acceptable capacitor voltage, which can be 1–20 cycles for electronic equipment. Fig. 6.8 shows the sag immunity for a PC obtained via laboratory testing.

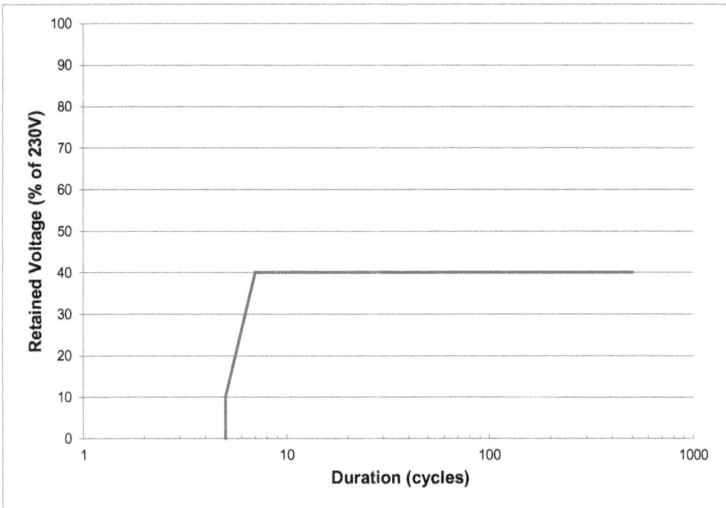

Fig. 6.8 Sag immunity for a PC.

6.6.1.2 *Variable speed drives*

Variable speed drives (VSD) have a similar front end to a SMPS. Their susceptibility has been given as lying in the range 85% voltage for ½ cycle to 70% voltage for 0.5 s [5]. Modern VSDs are manufactured with small sized smoothing capacitors to reduce volume and cost and they appear to be more sensitive to sags than data processing equipment. Shallow sags can allow continuous operation of the drive in a manufacturing plant but cause speed fluctuations which degrade the product. Unbalanced sags can cause a larger unbalance in the rectifier currents which can cause problems at smaller sag depths than for balanced sags. In addition to the inherent sag susceptibility of VSD based on component design, many VSDs also have in-built protection functions that will trip the device when a sag is detected. In some cases, this protection is much more sensitive than the underlying equipment capability. It is for this reason that VSDs are considered one of the most susceptible loads to voltage sags. Fig. 6.9 shows an example sag immunity curve for a VSD. In the case of this device, it is reasonably clear that the device has internal protection that causes a very fast trip if the input voltage falls below 80% of nominal.

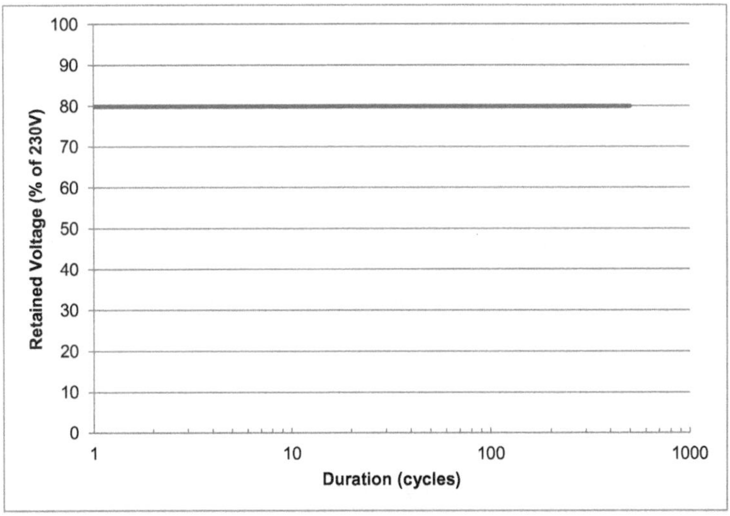

Fig. 6.9 Sag immunity for a VSD.

6.6.1.3 Induction motors

Directly connected induction motors will generally ride through sags but may depart from acceptable speed for sag durations of greater than 10 cycles. If the sag is severe enough and the load large enough, the motor may stall. Ultimately, the behaviour depends on the level of loading. When the sag recovers, the reduced motor speed causes a large inrush current which can slow down the voltage recovery.

6.6.1.4 AC contactors

AC contactors are known to be very sensitive to voltage sags, in some cases only tolerating 60% retained voltage for only a few cycles before tripping out. Their replacement by appropriately specified dc contactors is suggested as a means of improving plant sag performance. Fig. 6.10 shows a sag immunity curve for an AC contactor. In the case of this device, 0% retained voltage can only be tolerated for 3 cycles.

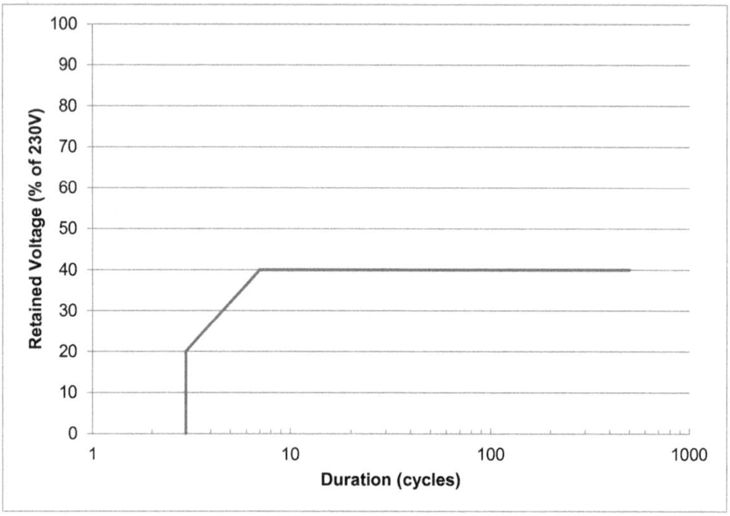

Fig. 6.10 Sag immunity for an AC contactor.

6.6.2 Industry sag immunity curves

6.6.2.1 CBEMA and ITIC curves

Whilst the previous section dealt with the performance of individual items of equipment, there have been a number of initiatives that have been undertaken in an attempt to provide an indication of the likely sag performance for a range of equipment. Probably the best known of these curves is the CBEMA curve (as shown in Fig. 6.11) and the successor to the CBEMA curve, the ITIC Curve (as shown in Fig. 6.12). Both of the CBEMA and ITIC curves are designed to illustrate the sag immunity if information technology equipment with CBEMA being an acronym for Computer and

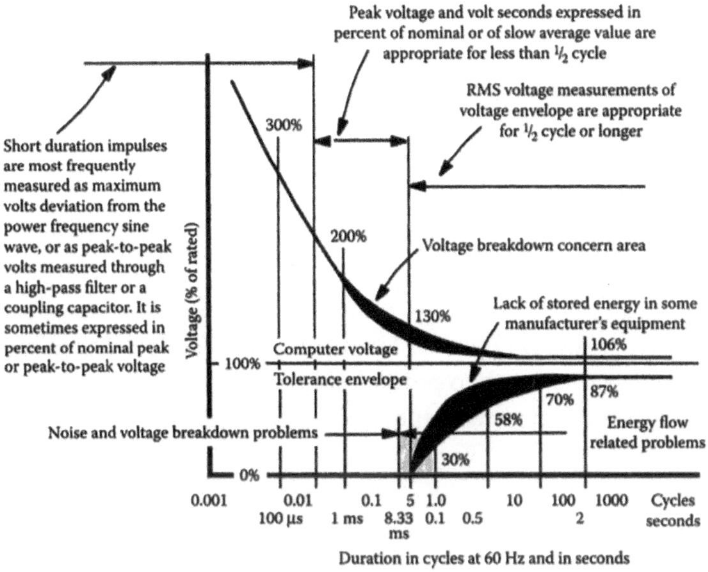

Fig. 6.11 Representation of the CBEMA curve.

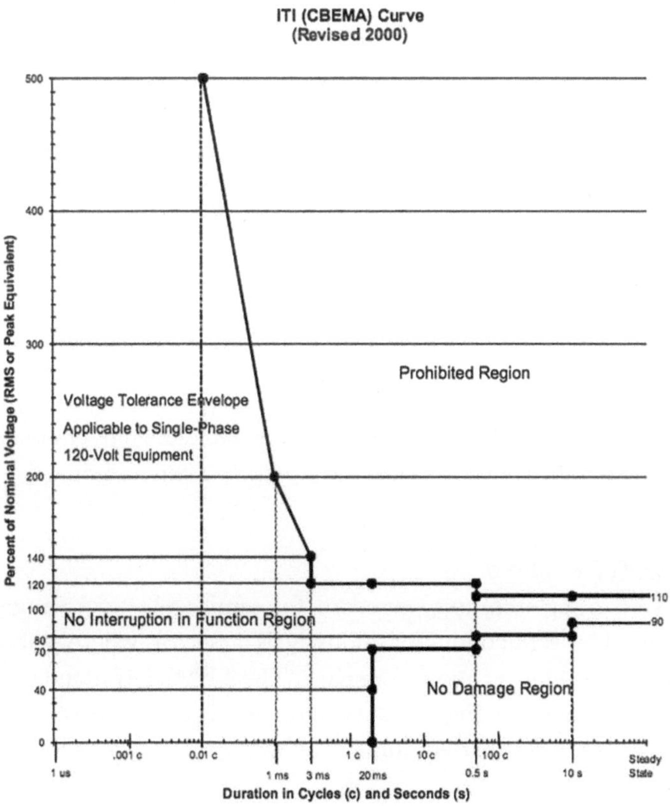

Fig. 6.12 ITIC curve.

Business Equipment Manufacturers Association and ITIC an acronym for Information Technology Industry Council.

It is relatively common to see either the CBEMA or ITIC curves built in to power quality monitoring software in order to generate graphics similar to the one shown in Fig. 6.13. Whilst these graphics can indicate whether or not information technology is likely to trip for any given sag, it is important to note that the curves provide no indication with respect to the capabilities of the electricity supply network. There is no known protection system that can clear faults fast enough to prevent voltage sags that are cannot be withstood (i.e. tolerated) by equipment which conforms with either the CBEMA or ITIC curves.

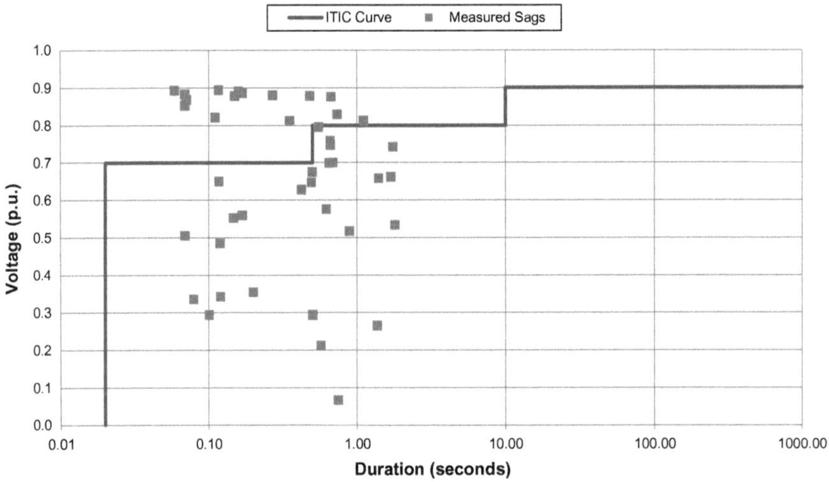

Fig. 6.13 Application of ITIC curve to indicate sags that will likely trip IT equipment (all sags to the lower right of the curve may be expected to cause trips or maloperation).

6.6.2.2 SEMI F47 curve

SEMI F47 is a sag immunity standard for semiconductor manufacturing equipment. The semiconductor industry is a sector where voltage sags can have a very high economic impact. The SEMI F47 standard defines the curve presented in Fig. 6.14 (where the x-axis is in 60-Hz cycles). The standard requires that equipment be immune to sags with depth and duration above the curve.

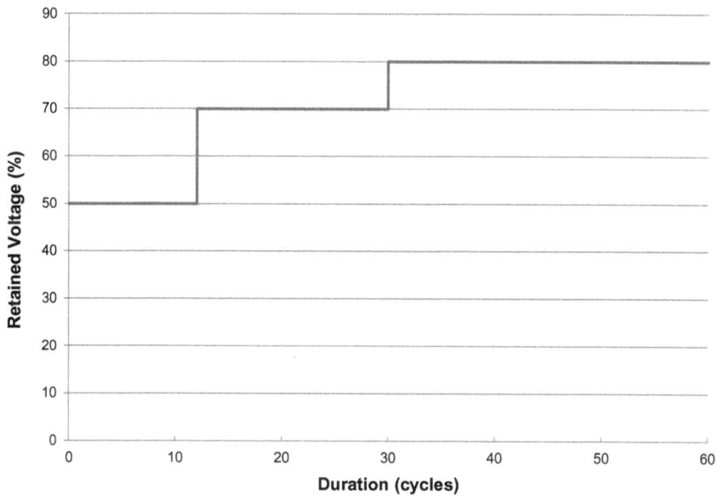

Fig. 6.14 SEMI F47 curve.

6.6.3 Network capability

Whilst the equipment curves shown in Section 6.6.2 can provide insight with respect to the sag immunity of individual pieces or groups of equipment type, they provide no indication of the likely performance of the network and whether or not the network performance is acceptable. For sags caused by network faults, the operation of the network to clear those faults (i.e. protection systems and operation) has a direct impact on the duration of the voltage sag. For faults in transmission networks or other areas with differential (unit) protection, faults will be cleared very quickly and sag durations may be restricted to a small number of cycles. However, the majority of faults will occur in distribution networks which typically use overcurrent protection as the primary mechanism for fault clearance. These protection schemes must grade with upstream setting and as such will have longer trip times. When examining whether or not network protection is clearing faults as quickly as it should, it may be useful to refer to a curve such as that shown in Fig. 6.15 which is reproduced from Ref. [6] and termed the Voltage Sag Protection Curve. This curve shows the expected clearing times for faults in networks with typical overcurrent protection settings. With respect to voltage sags, those with durations to the right of the protection curve may be an indication that protection systems are not clearing faults as quickly as might be expected. Examining the relationship between the protection curve and the CBEMA and ITIC curves, it can be seen that there are large areas below

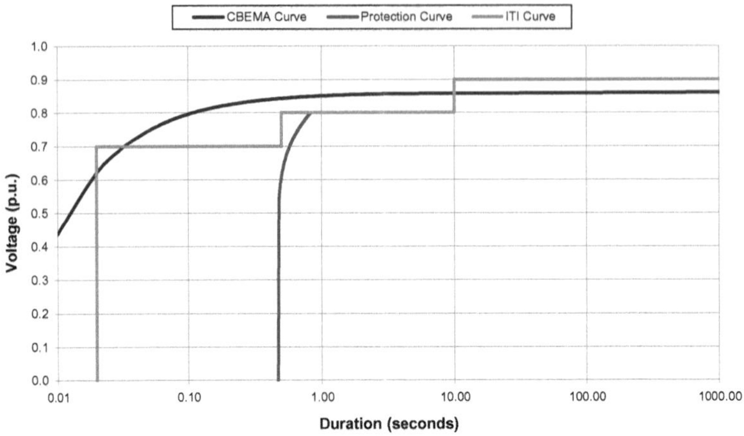

Fig. 6.15 Protection curve (and CBEMA and ITIC curves).

the CBEMA and ITIC curves which are to the left of the protection curve. In these regions, it is likely that equipment may trip; however, it is also likely that the network protection settings are unable to clear faults quickly enough to prevent the tripping.

6.6.4 Economic impact of voltage sags

There is a strong argument that can be made that voltage sags are the most costly of all power quality disturbances. Whilst perhaps not as costly as interruptions, voltage sags are much more prevalent than interruptions and if deep enough and/or long enough may have the same impact as a short supply interruption. Relatively shallow voltage sags can lead to the disruption of manufacturing processes due to equipment being unable to operate correctly at the reduced voltage levels. Industrial equipment such as variable speed drives and some control systems are particularly sensitive to voltage sags. In many manufacturing processes, loss of only a few vital pieces of equipment may lead to a full shut down of production leading to significant financial losses. For some processes which are thermally sensitive, a significant loss of material as well as the time taken to clean up and restart the process must also be considered.

There have been many studies which aim to quantify the cost of voltage sags. The results of these studies range from relatively modest cost associated with voltage sags through to very high costs generally at high technology industrial plants (such as semiconductor manufacturing). Table 6.2 reproduced from Ref. [7] presents the costs associated with voltage sags from a range of industries.

Table 6.3 reproduced from Ref. [8] presents another summary of the impact of voltage sags on various industries from the United States. The data

Table 6.2 Typical financial loss for voltage sags based on industry [7].

Industry	Typical financial loss per event (€)
Semiconductor production	3,800,000
Financial trading	6,000,000 per hour
Computer centre	750,000
Telecommunications	30,000 per minute
Steel works	350,000
Glass industry	250,000

Table 6.3 Impact of voltage sags on industry [8].

Industry	Loss per voltage sag ($US)
Paper manufacturing	30,000
Chemical industry	50,000
Automobile industry	75,000
Equipment manufacturing	100,000
Credit card processing	250,000
Semiconductor industry	2,500,000

presented agrees reasonably well with the data given in Table 6.2. It is stated in Ref. [8] that the cost to industry in the United States due to voltage disturbance is over $20 billion annually.

6.7 Mitigation of voltage sags

The large costs associated with voltage sags detailed in Section 6.6.4 can provide strong justification for the use of sag mitigation strategies and technologies. This section describes some of the most common methods of voltage sag mitigation, including theory of operation as well as advantages and disadvantages.

6.7.1 Network solutions

There are a range of network solutions that can be implemented to reduce the number and impact (i.e. depth and duration) of voltage sags. As with all mitigation of power quality disturbances, there is an economic evaluation that needs to be made which weighs up the benefits of mitigation against cost. Network solutions can be broadly classified into three main categories:

1. Fault reduction
2. Modification of fault clearing practices

3. Reduction of fault impact

Each of the earlier mentioned categories is examined in detail as follows.

6.7.1.1 Fault reduction

There are many causes of faults in electricity supply networks ranging from weather, through to equipment failure through to human or animal interaction with the power system. In all cases, the aim of fault reduction techniques is to minimise the exposure of the network to the mechanisms which cause faults. Less faults will ultimately mean less sags. The following are examples of fault reduction techniques:

- Vegetation management—contact between conductors and vegetation is a common cause of both transient and permanent network faults. Management of vegetation via activities such as pruning is generally a day-to-day maintenance activity for network operators. Most network operators will have vegetation management strategies in order to maintain network reliability and this in turn will reduce the prevalence of voltage sags.
- Use aerial bundled cable (ABC) or covered conductor—this reduces the exposure of the network to a range of fault causing mechanisms, including contact with vegetation or other airborne materials (during storms), direct lightning strikes and interaction with animals.
- Insulator washing—in areas of high pollution undesirable material can build up on insulators. This material reduces the dielectric strength of the insulator and can present a path for flashover and associated fault. Insulator washing can assist with maintaining insulator integrity.
- Animal guards—animal guards are designed to reduce the number of faults due to integration between animals and the power system.
- Shielding wires—these additional wires are usually only economical at voltages above 33 kV. The shielding wire is designed to protect the conductors from lighting strikes. The risk of flashover can be further reduced by decreasing the tower footing resistance.
- Undergrounding of conductors—placing conductors underground mitigates many sources of faults, including exposure to lighting and vegetation. However, this is an expensive solution and when a fault does occur it may be more difficult to locate and more costly to repair when compared to overhead networks.

6.7.1.2 Modification of fault clearing practices

Modification of fault clearing practices aims to either improve the sag characteristics (e.g. reduce sag duration) or reduce the number of customers exposed. Examples of actions that can be taken to modify fault clearing practices include the following:

- Increased sectionalising—this is achieved by placing line reclosers in series with the main feeder. Reclosers are used extensively to improve reliability by only isolating the sections of a feeder impacted by a fault to be opened. Reclosers can be graded to operate very quickly and clear downstream faults faster than may otherwise be possible using upstream protection. However, recloser operation which often incorporates a number of 'strikes' where the device will open and close in order to determine if a fault has cleared can increase the number of sags. In practical terms this does not necessarily significantly alter the impact of voltage sags as it is generally the first sag that will cause equipment to trip or maloperate.
- Revision of protection setting—as discussed in Section 6.6.3, which detailed the protection curve, network protection settings will have significant impact on sag characteristics. There may be opportunities to refine protection settings or instal alternate protection schemes to improve sag performance.

6.7.1.3 Reduction of fault impacts

These techniques aim to reduce the fault current which in turn will reduce the depth of the voltage sag. Examples of techniques include

- Neutral earth resistors or reactors—these are connected to the neutral point of zone transformers. Most faults are line-ground or earth faults. Neutral earthing resistors increase the impedance in the earth path and as such reduce the fault current. Care must be taken, however, as the neutral voltage will rise during earth faults and insulation grading and lighting arrester ratings must be adjusted to account for this.
- Series reactors—these can be added to each feeder and will increase fault impedance. However, in their simplest applications, series reactors may induce other problems related to the normal load current such as voltage regulation. Flicker and harmonics magnitudes may also increase. One potential solution to this problem is superconducting fault current limiters. These series devices present a low impedance for normal operating current but a much higher impedance when subject to fault current.
- Resonant earthing—also known as arc suppression coils or Petersen coils. This mitigation technique involves connection of a special adjustable iron cored reactor in the earth path. This is a special form of neutral-reactance grounding. If properly applied, arc suppression coils can reduce the fault current to zero and also prevent any arcing (which is advantageous in high bushfire risk locations).

6.7.2 Coil hold-in devices

Contactor coils are devices which have traditionally been susceptible to voltage sags. In some cases, the loss of a single contactor can lead to the loss of a whole production line even if all of the other equipment is not susceptible to the voltage sag. In some cases, a change to the contractor circuit or type can be a very simple and cost-effective method of voltage sag mitigation. Coil hold-in devices are connected between the AC supply and the contactor and can generally allow a contactor to remain energised for voltage sags down to 25% retained voltage.

6.7.3 Ferroresonant transformer

Ferroresonant transformers, also known as a constant voltage transformer (CVT), which are fully described in Chapter 2 in their application for mitigating steady-state voltage variation can also be useful for mitigating relatively shallow voltage sags.

6.7.4 Uninterruptible power supply

Uninterruptible power supplies (UPS) mitigate voltage sags by supplying the load using stored energy. Upon detection of a voltage sag, the load is transferred from the mains supply to the UPS. Obviously, the capacity of load that can be supplied and the duration for which it can be supported are related to the rating of the UPS and the amount of energy storage available. UPS systems have the advantage that they can mitigate all voltage sags, including outages for significant periods of time (depending on the size of the UPS).

There are 2 main topologies of UPS in use: on-line and off-line. These topologies differ in the way that they supply the load. Fig. 6.16 shows a schematic of an off-line UPS whilst Fig. 6.17 shows a schematic of an on-line UPS. Comparison of the figures shows that the difference between the two systems is that for an on-line UPS the load is always supplied by the UPS, whilst for off-line systems, the load is transferred from the mains supply to the UPS by a static changeover switch upon detection of a voltage sag. The lack of a changeover switch renders the on-line system more reliable as any failure of the changeover switch will result in the off-line UPS being ineffective. A disadvantage of UPS systems is that the energy storage components (mostly batteries) must be maintained and replaced periodically.

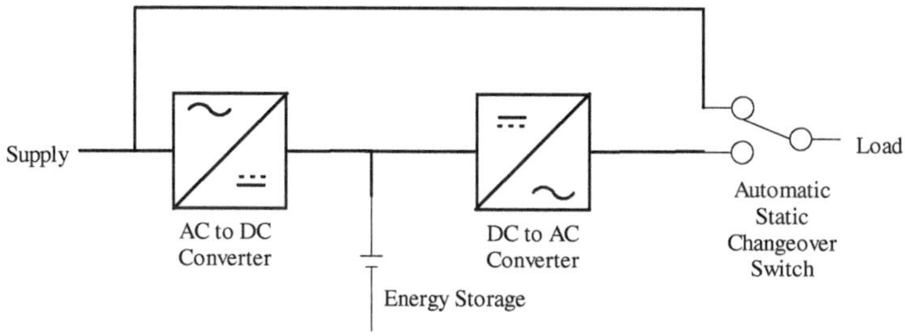

Fig. 6.16 Block diagram of an off-line UPS.

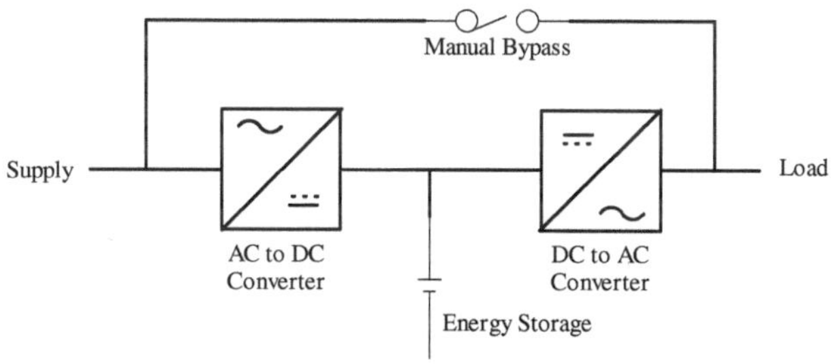

Fig. 6.17 Block diagram of an on-line UPS.

Small UPS systems are relatively simple and cheap. However, large units are complex and highly expensive due to the need for large energy storage capacities.

6.7.5 Flywheel and motor–generator (MG)

Flywheel systems use the energy stored in the inertia of a rotating flywheel to mitigate voltage sags. In the most basic system, a flywheel is coupled in series with a motor and a generator which in turn is connected in series with the load. The flywheel is accelerated to a very high speed and when a voltage sag occurs, the rotational energy of the decelerating flywheel is utilised to supply the load. Flywheel storage systems are effective for mitigation of all voltage sags including interruptions and can supply the load for a significant period of time (up to several seconds depending on the size of the flywheel). Fig. 6.18 shows the basic principle of the flywheel and motor–generator.

Flywheels have maintenance and reliability advantages over other energy storage systems such as batteries. However, if large energy storage capacities are required, flywheels must be large and heavy. Further, the configuration

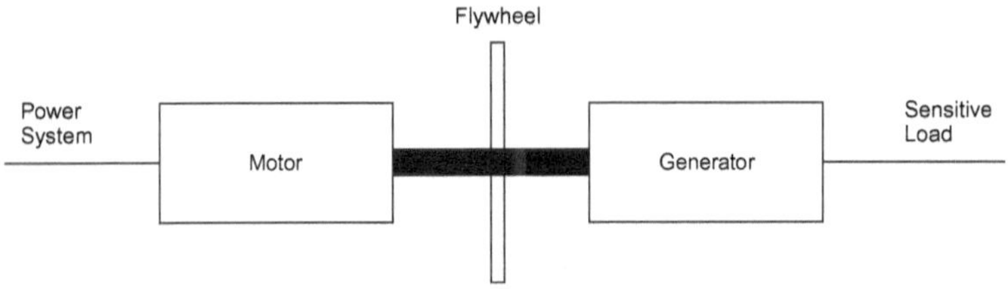

Fig. 6.18 Basic flywheel motor–generator configuration [4].

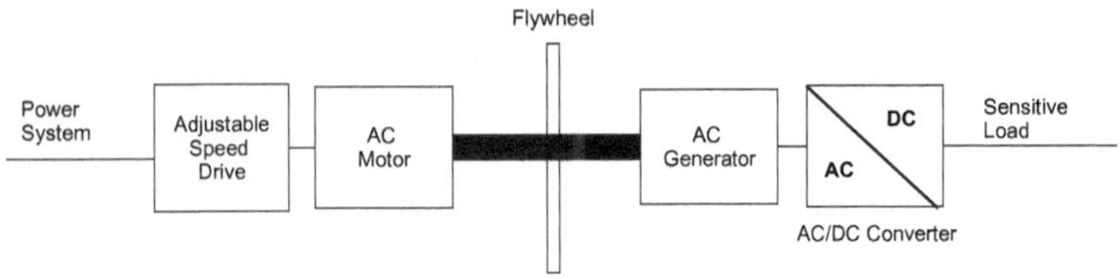

Fig. 6.19 Advanced flywheel motor–generator configuration [4].

shown in Fig. 6.18 will have high losses during normal operation [4]. A number of solutions have been proposed to overcome this issue and most involve the inclusion of power electronics into the system. Such a solution is presented in Fig. 6.19. In this configuration, the motor which drives the flywheel is connected through a variable speed drive. This connection arrangement results in better starting characteristics for the flywheel and efficiency gains for the motor. Connection of the ac generator to a voltage source converter as shown increases the amount of energy that can be extracted from the flywheel due to the fact that the converter is able to produce a constant DC voltage, which may then be used directly or converted back to AC voltage, over a wide speed range.

6.7.6 Dynamic voltage restorer

Dynamic voltage restorers (DVR) are complicated static devices which work by adding the 'missing' voltage during a voltage sag. Basically this means that the device injects voltage into the system in order to bring the voltage back up to the level required by the load. Injection of voltage is achieved by a switching system coupled with a transformer which is connected in series with the load. There are two types of DVRs available: those with and without energy storage. Devices without energy storage are able to correct the voltage waveform by drawing additional current from the

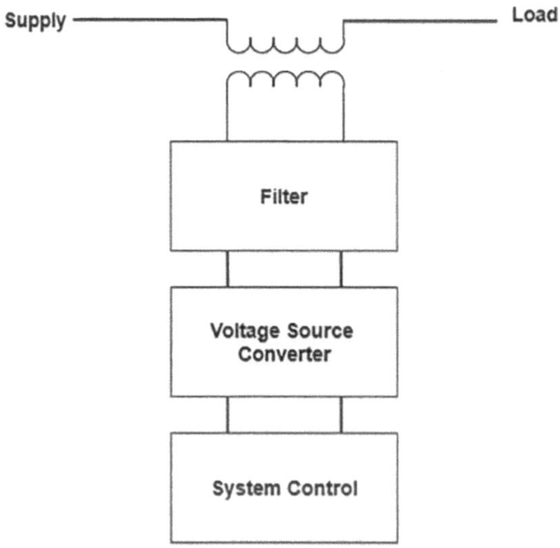

Fig. 6.20 Block diagram of a DVR.

supply. Devices with energy storage use the stored energy to correct the voltage waveform. The difference between a DVR with storage and a UPS is that the DVR only supplies the part of the waveform that has been reduced due to the voltage sag, not the whole waveform. In addition, DVRs generally cannot operate during interruptions. Fig. 6.20 shows a schematic of a DVR. As can be seen, the basic DVR consists of an injection/booster transformer, a harmonic filter, a voltage source converter (VSC) and a control system. For readers who are interested in further knowledge of DVR systems, the article in Ref. [9] gives a thorough description of the design and operation of DVRs.

DVR systems have the advantage that they are highly efficient and fast acting. It is claimed in Ref. [9] that the DVR is the most economic solution for mitigating voltage sags based on its size and capabilities. In the case of systems without storage, none of the inherent issues with storage are relevant. Another advantage of DVR systems is that they can be used for purposes other than just voltage sag mitigation. These added features include harmonic mitigation, fault current limiting, power factor correction and reduction of transients.

6.7.7 Static var compensator

A static var compensator (SVC) is a shunt-connected power electronics-based device which works by injecting reactive current into the load, thereby supporting the voltage and mitigating the voltage sag. SVCs may

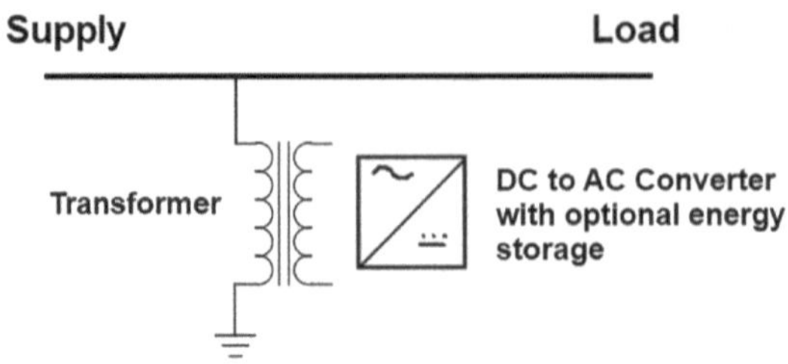

Fig. 6.21 Block diagram of a SVC.

or may not include energy storage with those systems which include storage being capable of mitigating deeper and longer voltage sags. Fig. 6.21 shows a block diagram of a SVC.

6.7.8 Sag proofing transformers

Sag proofing transformers, also known as voltage sag compensators, are basically a multiwinding transformer connected in series with the load. These devices use static switches to change the transformer turns ratio to compensate for the voltage sag. Sag proofing transformers are effective for voltage sags to approximately 40% retained voltage. Fig. 6.22 shows a block diagram of a sag proofing transformer.

Sag proofing transformers have the advantage of being basically maintenance free and do not have the problems associated with energy storage

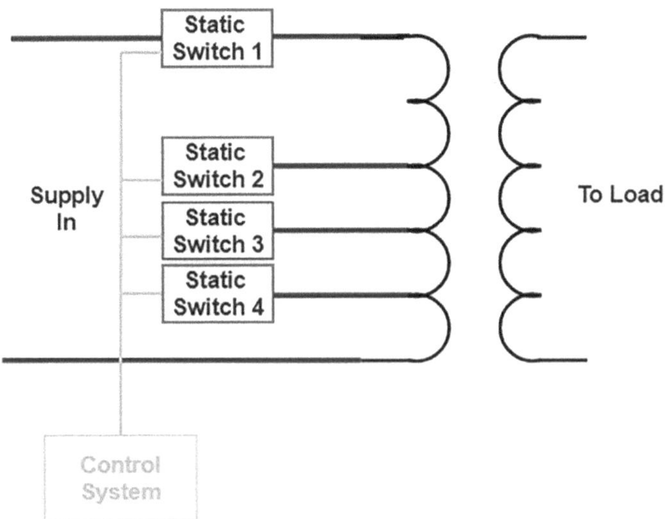

Fig. 6.22 Schematic of a sag proofing transformer.

components. A disadvantage is that at present, sag proofing transformers are only available for relatively small loads of up to approximately 5 kVA. With the transformer connected in series, the system also adds to losses and any failure of the transformer will lead to an immediate loss of supply.

6.7.9 Static transfer switch

For facilities with a dual supply, one possible method of voltage sag mitigation is through the use of a static transfer switch (STS). Upon detection of a voltage sag, these devices can transfer the load from the normal supply feeder to the alternative supply feeder within ½ a cycle. The effectiveness of this switching operation is highly dependent on how independent of each other the 2 supply feeders are and the location of the event leading to the voltage sag. Ideally, with a dual feeder supply, the 2 feeders should be supplied by different substations. Obviously, there are significant costs associated with dual supplies even if they are available.

6.7.10 Motor soft starting

The large inrush currents associated with energisation of direct on-line motors are a well-known source of voltage sags. There are a number of technologies which can be used to mitigate inrush currents and as such reduce or eliminate the associated voltage sags. These 'soft-starting' technologies include

- Star-delta starters—these devices perform a motor start in a star configuration which has a lower voltage (line-neutral voltage as opposed to line–line, i.e. a factor of $\sqrt{3}$ lower). With reduced voltage the motor will draw a reduced current. As the motor speeds up, the starter switches from star configuration to the delta configuration. The disadvantage of star-delta starters is that the reduced voltage start results in reduced starting torque.
- Electronic soft starters—these devices use electronics (typically SCRs) to control the voltage to the motor during the starting phase. Voltage control is achieved through phase angle control. Electronic soft starters can reduce inrush current; however, they will draw highly distorted current during the starting phase and are not as effective as variable speed drives (but are likely much cheaper).
- Variable speed drives—these power electronic devices are able to eliminate the sags due to motor energisation and can also provide benefits for energy efficiency.

6.7.11 Energy storage technologies

Although not sag mitigations device by themselves, energy storage systems are essential to most of the above sag mitigation technologies. As such, a short description of energy storage technologies is relevant. In cases of devices such as the DVR and SVC, the device is compatible for use with a number of different energy storage technologies. The choice of technology generally depends on the application, maintenance requirements and cost. At present there are four main energy storage technologies that may be applied to sag mitigation technologies. These are flywheel, batteries, superconducting magnetic energy storage (SMES) and capacitors.

6.7.11.1 Flywheels

Flywheel energy storage systems are one of the oldest storage technologies with examples dating back to the 11th century [10]. Modern flywheel systems incorporate advanced materials such as carbon fibre, have magnetic bearings and may spin in a vacuum to reduce losses.

Flywheels have the advantage that they are simple and low maintenance. They also have a long lifespan. Generally flywheels do not contain materials which are particularly dangerous to the environment. On the downside, they introduce losses into the system and may not charge as fast as other devices such as capacitors.

6.7.11.2 Batteries

Battery energy storage is another system that has been in existence for a considerable period of time. Although an area of continual research, battery technologies are well developed and well understood. Batteries are relatively cheap and when maintained correctly provide excellent performance. In addition, batteries have the highest energy density of all the considered energy storage technologies. The main disadvantage of batteries is that they have a finite number of charge cycles and hence a limited lifespan. They also contain materials which may be hazardous to the environment.

6.7.11.3 Capacitors

Capacitors and the modern super or ultra-capacitors are becoming a more popular choice for energy storage. Capacitors are simple and have very fast charge times. They do not have the charge cycle limitations of batteries and hence may have a longer lifespan if not subjected to overvoltage stress. Disadvantages of capacitors include relatively higher costs compared to batteries and relatively lower energy density levels compared to batteries.

6.7.11.4 *Superconducting magnetic energy storage*

SMES systems are a developing technology which utilises the properties of superconducting material to store energy in magnetic fields. SMES systems have very fast charge and discharge times which make them an attractive energy storage system for sag mitigation. Another advantage of SMES systems is the very low losses due to the superconducting characteristics. The main disadvantage of SMES over batteries at present is the cost. SMES systems also have all of the disadvantages associated with superconducting technology, not least of which is the need for liquid nitrogen to maintain the cryogenic temperatures required for superconductivity.

6.7.12 Cost of sag mitigation technologies

6.7.12.1 *Cost of mitigation technologies*

This section attempts to quantify and compare the costs of the various mitigation devices discussed previously. For all technologies, there will be two costs involved. The first is the initial purchase price of the equipment whilst the second is the maintenance costs associated with the selected equipment. There are a number of studies which give the costs of mitigation technologies and not all agree well with each other. Table 6.4 presents a range of costs for a number of the mitigation technologies discussed before. It can be seen that the DVR is the cheapest mitigation technology on a cost per kVA basis. However, DVR systems are usually only used for large loads and the costing is based on this fact. UPS systems or ferroresonant transformers are the only viable mitigation strategies for small loads.

Table 6.4 Cost of voltage sag mitigation technologies.

Mitigation technology	Initial cost ($US)	Operation cost (% of initial cost per year)
Coil hold-in devices	100–150 each [11]	N/A
Ferroresonant transformer (CVT)	1000/kVA [12]	1 [12]
UPS	500/kVA [12]–1000/KVA [11]	1.5–2.5 [12] 10 [13]
Flywheel	500/kVA [12]	0.7 [12]
DVR (50% voltage boost)	250/kVA [12]	0.5 [12]
Statcom	400/kVA [13]	5 [13]
Static switch (10 MVA)	600,000 [12]	0.5 [12]

Table 6.5 Comparison of the cost of storage technologies.

		Costs of energy storage ($)		
Power	**Ride-through time**	**SMES**	**BESS**	**Capacitors**
300 kW	1 s	183,000	6300	56,000
	60 s	389,000	6300	3 350,000
3 MW	1 s	411,000	63,000	558,000
	60 s	1,064,000	63,000	33,500,000

6.7.12.2 Comparison of costs of storage technologies

Table 6.5, reproduced from Ref. [4], presents a comparison of the cost of energy storage technologies depending on the application. It can be seen that for all applications, battery energy storage systems (BESS) remain the cheapest solution whilst, depending on the application, capacitors or SMES may be the next cheapest.

6.8 Assessment and reporting of voltage sags

There have been many methods suggested for assessment and reporting of voltage sags. However, to date, there is no prescriptive methodology that has achieved international consensus. This section provides a summary of a range of the proposed voltage sag indices and reporting methodologies.

6.8.1 Phase and time aggregation

For the purposes of analysis and reporting, it is often useful to reduce multiphase sags or sags that occur over a short time period that may be due to the same root case, e.g. single-phase faults developing into multiphase faults or multiple sags due to recloser operations, to a single depth and duration value. This process is known as time and phase aggregation and details are provided as follows:

1. **Phase aggregation** involves combining sags that occur at the same timestamp across multiple phases. There are a number of methods of achieving phase aggregation with two examples given as follows:
   - The depth of the sag on each phase can be assessed and the characteristics of the sag with the deepest depth, i.e. least retained voltage, used to reduce the multiphase values to a single sag.
   - An individual sag index which provides assessment based on both depth and duration can be calculated for the voltage sag on each phase. The retained sag data (timestamp, residual voltage and

duration) is the data which corresponds to the sag with the largest sag index value. This method may be considered superior to other methods of phase aggregation as it recognises the impact of both depth and duration, e.g. a long, relatively shallow sag may have the same or greater impact on equipment as a short deep sag.

- **Time aggregation** involves combining sags that occur close together in time. The rationale here is that sags which are temporally close are likely due to the same root cause and/or may be a result of recloser operation. Time aggregation is achieved by applying a fixed time window (often 1 min) beginning from the timestamp of the first sag. Time aggregation is applied after phase aggregation. In many cases, the retained sag characteristics are those for the most severe sag during the fixed time window.

6.8.2 Indices for individual sag events

There are many indices which have been considered for voltage sags in the literature. No firm consensus has been reached on appropriate indices and even standards generally give a number of options as opposed to a firm recommendation for a single sag index. It is impossible to include a thorough literature review of all of the published sag indices. Instead, a summary of some of the most significant contributions is given as follows.

IEEE standard 1564-2014 [14] collates a large number of potential single event and site indices for voltage sags. However, no specific recommendation is given for a single event or site voltage sag site index. The single sag event indices cited in IEEE 1564 are as follows:

- Voltage sag energy index—"The voltage sag energy is the duration of an interruption that would result in the same loss of energy for an impedance load" [14].
- Voltage sag severity index—"The voltage sag severity is calculated from the retained voltage in per unit and the duration of a voltage sag in combination with a reference curve" [14]. This index is calculated by comparing voltage sag characteristics (duration and retained voltage) with a reference curve (e.g. CBEMA or SEMI F47). Voltage sags with characteristics equal to the curve have an index of one. Values below the curve have increasing index values whilst those above the curve have decreasing values.

A number of sag indices have been suggested by the University of Wollongong for use in wide-scale power quality monitoring surveys. The first index is the Sag SAIFI as described in Ref. [15]. This index relates sag performance with the reliability measure SAIFI. Sag SAIFI is similar in practice to the sag severity index given in IEEE 1564 and uses a reference curve based on equipment performance testing. In this way, the Sag SAIFI equates voltage sag events into equivalent interruptions based on actual equipment performance. For any given voltage sag, the Sag SAIFI value varies from 0 to 1. A Sag SAIFI value of 1 indicates that all equipment would be expected to be tripped by the voltage sag. Fig. 6.23 shows sag SAIFI contours on a voltage time plane. Each individual voltage sag can be assigned a value depending on where it occurs on the voltage–time plane based on depth and duration.

The second novel index developed by the University of Wollongong is the protection index. This index utilises the protection curve illustrated in Fig. 6.15 to relate sag performance to typical distribution level nominal voltage overcurrent protection settings. The protection index is calculated as follows:

For a sag with retained voltage V (noting that sags with retained voltage greater than 90% and less than 10% are omitted), the associated clearing time (T_p) according to the Protection curve is calculated as follows:

$$T_p = 2.7V^9 + 0.47$$

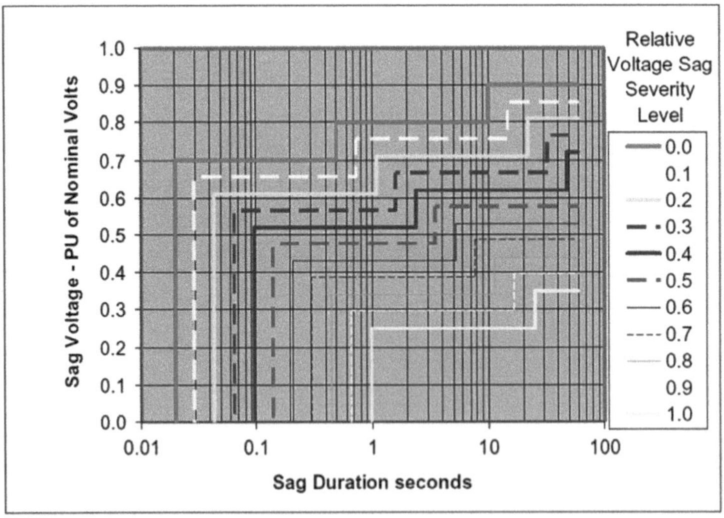

Fig. 6.23 Sag SAIFI contours.

The duration T of the sag is then compared to T_p as follows:

- If $T < T_p$

$$\text{Protection Index} = 0$$

- If $T > T_p$

$$\text{Protection Index} = T - T_p$$

6.8.3 Site reporting

Site reporting involves calculation of a single index or use of another method to describe the performance of an individual site over a given period of time. Methods of reporting site performance for sags can be divided into methods that calculate site indices and methods that use tabular reporting formats to characterise site performance.

6.8.3.1 Site indices

SARFI (System Average RMS Variation Frequency Index) is a well-known index for characterising sag performance at a site. SARFI provides a count or rate of voltage sags, swells and/or interruptions for a system. There are two versions of SARFI: SARFI-X and SARFI-Curve. "SARFI-X corresponds to a count or rate of voltage sags, interruptions, and/or swells below or above a specified voltage threshold. For example, the value of SARFI-90 would be a count of all the voltage sags and interruptions that are below 0.90 per unit, or 90% of the reference voltage." "SARFI-Curve corresponds to a rate of voltage sags below an equipment compatibility curve. For example, SARFI-CBEMA considers voltage sags and interruptions that are below the lower Computer Business Equipment Manufacturers Association (CBEMA) curve" [14].

A site index can also be determined using both the Sag SAIFI and the Protection Indices as described before. In both cases, the site index is simply the sum of the individual indices over the reporting period.

6.8.3.2 Tabular reporting formats

A number of tabular formats have been proposed in order to report sag behaviour at a site. These tabular formats generally provide counts of the number of sags based on depth and duration characteristics which correspond to one of the table cells. Limits can be placed on the number of sags in each cell in order to provide assessment of site compliance.

To date, no single tabular reporting format has been settled on by the international community. Many of the tabular reporting formats have deficiencies as they can be complex to apply and do not readily relate dip performance to either the impact on equipment connected to the network or network capability. A number of the most common tabular formats are detailed as follows.

The IEC Technical Report IEC/TR 61000.2.8:2009 [16] describes the electromagnetic disturbance phenomena of voltage dips and short interruptions in terms of their sources, effects, remedial measures, methods of measurement and measurement results. The standard recommends a tabular approach to reporting of voltage sag data and identifies the UNIPEDE Disdip tabular format as defined in Ref. [17], as reproduced, in Table 6.6 as an appropriate sag reporting methodology.

EN50160 [18] also specifies a tabular method of characterising voltage dips that is similar to the UNIPEDE Disdip table but with variation in the cell values. The tabular classification method given in EN50160 is reproduced in Table 6.7.

The South African network operator ESKOM has developed a 'Voltage Dip Window' [19] method for reporting and assessment of sags. This methodology is somewhat simplified method of tabular reporting compared to the EN50160 or UNIPEDE tables. The Voltage Dip Window (presented in Fig. 6.24) divides the voltage time plane up into 5 segments with a letter for each segment.

6.8.4 Assessment and reporting of overall network performance

Assessment of overall network performance requires a method of combining site reports in order to provide insight into whole of network outcomes. Tabular methods similar to those described in Section 6.8.3.2 can be used for assessment and reporting of overall network performance. In the case where tabular methods are used to report whole of network performance, the table cells are populated with the number of sags corresponding to each table cell across the entire network as opposed to the individual site.

Sag indices can also be combined to allow for whole of network assessment and reporting. In the case of an index such as SARFI, the count of sags below the nominated metric becomes a count across and entire network as opposed to a single site. For indices such as Sag SAIFI and the protection index, statistical measures can be applied to the individual site indices.

Table 6.6 UNIPEDE Disdip tabular format for voltage sag data.

Retained voltage u (% of reference voltage)	Duration t (s)						
	$0.01 < t \leq 0.02$	$0.02 < t \leq 0.1$	$0.1 < t \leq 0.5$	$1 < t \leq 3$	$3 < t \leq 20$	$20 < t \leq 60$	$60 < t \leq 180$
$90 > u \geq 85$							
$85 > u \geq 70$							
$70 > u \geq 40$							
$40 > u \geq 10$							
$10 > u \geq 0$							

Reproduced from Disdip Group, Voltage dips and short interruptions in medium voltage public electricity supply systems, Report Prepared for International Union of Producers and Distributors of Electrical Energy (UNIPEDE), 1990.

Table 6.7 EN50160 voltage dip classification table [18].

Residual voltage (u) %	Duration t ms				
	$10 \leq t \leq 200$	$200 < t \leq 500$	$500 < t \leq 1000$	$1000 < t \leq 5000$	$5000 < t \leq 60{,}000$
$90 > u \geq 80$	CELL A1	CELL A2	CELL A3	CELL A4	CELL A5
$80 > u \geq 70$	CELL B1	CELL B2	CELL B3	CELL B4	CELL B5
$70 > u \geq 40$	CELL C1	CELL C2	CELL C3	CELL C4	CELL C5
$40 > u \geq 5$	CELL D1	CELL D2	CELL D3	CELL D4	CELL D5
$5 > u$	CELL X1	CELL X2	CELL X3	CELL X4	CELL X5

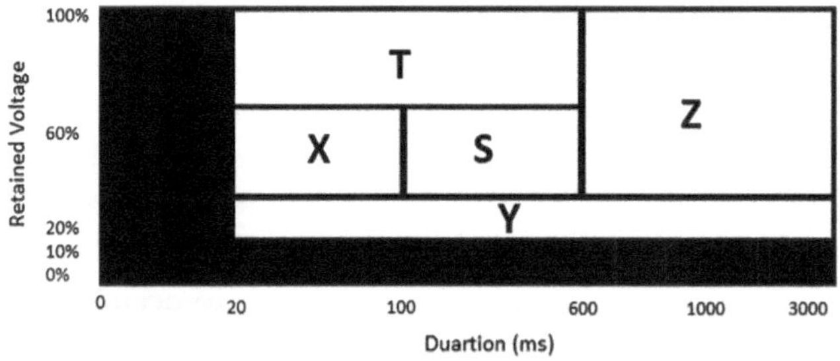

Fig. 6.24 ESKOM voltage dip window.

For example, the network value for Sag SAIFI may be selected to be the 95th percentile value of all SAG SAIFI values across all sites.

In Section 6.6.2 the disconnect between equipment sag immunity and practical network performance (i.e. the ability of the network to clear the faults that cause voltage sags) was identified. In order to address this disconnect, a number of European organisations have introduced the concept of 'responsibility sharing' [20]. Under this arrangement, curves are specified in the voltage–time plane which subdivides the plane into areas of responsibility for equipment manufacturers and network operators. A generic example of such a curve, adapted from Ref. [20], is shown in Fig. 6.25. In this figure, the voltage–time plane has been divided into three regions: A, B and C. The region defined by Area A is a region of depth and duration where network protection cannot generally operate to clear faults fast enough to prevent sags with these characteristics. The responsibility sharing concept acknowledges that it is not practical for the network operator to prevent sags in this region, and as such, responsibility lies with the installation

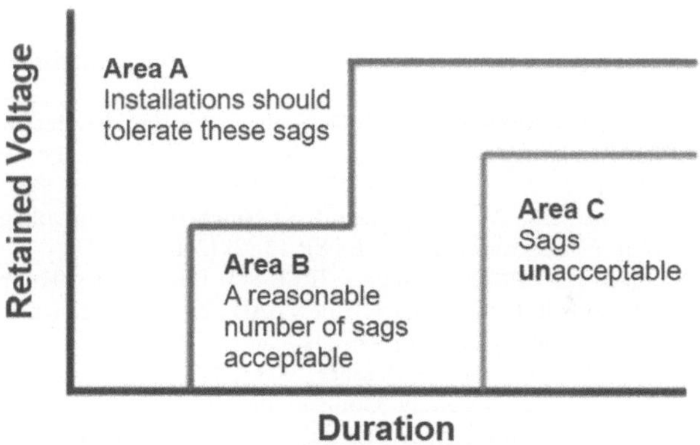

Fig. 6.25 Example of generic dip responsibility sharing curves [20].

or equipment owner. Region B is an area of shared responsibility acknowledging that network operators may be able to clear some but not all of the faults which would cause sags in this region. As such, Area B defines an area where the network operator must limit the number of dips but the installation must also be aware that they can occur and can choose to take action to mitigate them. The area of the plane defined by Area C is a region where network operators are not permitted to allow dips to occur. In this region, it is the protection settings and other network design features which are controlled by the network operator that will define the dip characteristics.

References

[1] International Electrotechnical Vocabulary, Chapter 604: Generation, Transmission and Distribution of Electricity—Operation, 1998.

[2] IEEE, IEEE Recommended Practice for Monitoring Power Quality, 2009.

[3] S. Elphick, V. Gosbell, V. Smith, R. Barr, The Australian long term power quality survey project update, in: 14th International Conference on Harmonics and Quality of Power, ICHQP'10, Bergamo, Italy, 26–29 September 2010, 2010.

[4] M.H.J. Bollen, Understanding Power Quality Problems Voltage Sags and Interruptions, John Wiley & Sons, New Jersey, 2000.

[5] IEEE Project 1346 Working Group, Electric Power System Compatibility with Industrial Process Equipment Part 1: Voltage Sags, Proceedings of Industrial and Commercial Power Systems Conference, 1–5 May 1994, 1994, pp. 261–266.

[6] R.A. Barr, V.J. Gosbell, S. Perera, The voltage sag protection curve, in: 12th International Conference on Harmonics and Quality of Power, ICHQP2006, Cascais, Portugal, 1–5 October, 2006, 2006.

[7] D. Chapman, Power Quality Application Guide—The Cost of Poor Power Quality, 2001.

[8] B.H. Chowdhury, Power quality, IEEE Potentials 20 (2) (2001) 5–11.

[9] C. Benachaiba, B. Ferdi, Voltage quality improvement using DVR, Electr. Power Qual. Util. J. XIV (1) (2008) 39–45.

[10] Wikipedia, Flywheel, 2012, Webpage, Last Accessed 8 February 2012, Available: http://en.wikipedia.org/wiki/Flywheel#History.

[11] Pacific Gas and Electric Company, Voltage Sag Ride-Through Mitigation in Sequence by Increasing Cost, 1999.

[12] T. Thasanutariya, S. Chatratana, M. McGranaghan, Economic evaluation of solution alternatives for voltage sags and momentary interruptions, Electr. Power Qual. Util. Mag. 1 (2) (2005) 17–26.

[13] M. Didden, R. Belmans, W. D'Haeseleer, Cost-benefit analysis of voltage sag mitigation methods in fiber extrusion plants, ETEP 13 (2) (2003) 73–77.

[14] IEEE, IEEE guide for voltage sag indices, IEEE Std. 1564-2014, 2014, pp. 1–59.

[15] R.A. Barr, V.J. Gosbell, I. McMichael, A new SAIFI based voltage sag index, in: 13th International Conference on Harmonics and Quality of Power, ICHQP2008, Wollongong, Australia, 28 September – 1 October 2008, 2008.

[16] Standards Australia, Electromagnetic Compatibility (EMC) Part 2.8: Environment—Voltage Dips and Short Interruptions on Public Electric Power Supply Systems With Statistical Measurement Results, AS/NZS IEC/TR 61000.2.8:2009, 2009.

[17] Disdip Group, Voltage dips and short interruptions in medium voltage public electricity supply systems, Report Prepared for International Union of Producers and Distributors of Electrical Energy (UNIPEDE), 1990.

[18] CENELEC, Voltage Characteristics of Electricity Supplied by Public Electricity Networks, EN 50160:2010, 2010.

[19] ESKOM, South African Power Quality Standard NRS 048-2: 1996, 1996.

[20] M.H.J. Bollen, J. Esteves, K. Brekke, K. Niall, M. Delfanti, Voltage quality monitoring, dips classification and responsibility sharing, in: Harmonics and Quality of Power (ICHQP), 2012 IEEE 15th International Conference on, 17–20 June 2012, 2012, pp. 926–931.

CHAPTER 7

Implications of equipment behaviour on power quality

7.1 Introduction

The vast majority power quality disturbances occur as a result of the interaction between the current flowing in an electricity supply network and the impedance of the network. The magnitude of variations such as steady-state voltage, harmonic distortion, voltage unbalance and flicker is directly related to the interaction between this load current and the network impedance. For rapid voltage change and voltage sags, it is the interaction between either the load current or the fault current and the network impedance that determines the voltage change or residual voltage and hence one aspect of the severity of the voltage fluctuation or sag.

In this chapter the implications of the operation of a range of common equipment types on power quality are described. In the first instance, a technical description of the operating characteristics of common loads which are connected to electricity distribution systems is provided. Subsequently, a description and the characteristics of common power system loads are provided and where applicable this includes waveforms and analysis of harmonic distortion spectra.

An overview of the impacts of variation in the characteristics of the supply voltage on equipment performance has been provided in order to provide the reader with an indication of the implications of supply voltage characteristics on power quality performance. The chapter concludes by providing a summary of standards designed to manage the power quality emissions of power system loads.

7.2 Power electronic converters

This section describes the typical power electronic technologies that are used to convert the supply voltage (sinusoidal 50- or 60-Hz AC) to the magnitudes and frequencies required for many types of equipment to operate. Many of the loads (appliances) examined later in the chapter rely on one

Applied Power Quality
https://doi.org/10.1016/B978-0-323-85467-2.00002-0

(or more) of these technologies to operate. The following conversions are considered:

- AC/AC conversion (phase control)
- AC/DC conversion (rectifier)
- DC/AC conversion (inverter)
- Complex conversion circuits such as variable speed motor drives which are AC/DC then DC/AC conversions

By their very nature, power electronics work by altering the input voltage waveform. The converter designs are reasonably complex electrical circuits containing nonlinear components. As such, power electronic converters are a well-known source of waveform distortion. Various circuit design methodologies can be used to mitigate the harmonic emissions of power electronic converters; however, the extent and efficacy of this mitigation often depends on the size and costs of the device, i.e. effort might be made to mitigate the amount of waveform distortion from a high power, high cost load such as a large solar inverter or variable speed drive but little attention is generally focused on low power, low cost devices such as domestic LED lighting.

7.2.1 Phase control AC/AC conversion

Phase control AC/AC conversion is generally applied in situations where a reduction in AC voltage is desirable (e.g. dimming of incandescent lighting). The basic component of this type of converter is the AC Static Switch. This static switch consists of two Silicon Controlled Rectifiers (SCRs) connected 'back-to-back' as shown in Fig. 7.1.

The AC static switch can be used to make a connection in less than 1 ms. The switch then disconnects at first current zero. A simple example of the application of this device is given in Fig. 7.2 for a resistive load.

This technology is called phase control as the output voltage of AC/AC converter is controlled by applying gate drive pulses to the gate terminals of S1 and S2 which are synchronised to the fundamental frequency of the mains

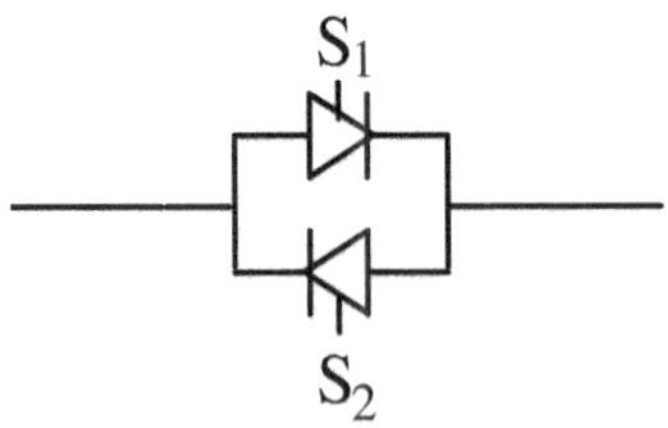

Fig. 7.1 Diagram of AC static switch. *(From V.J. Gosbell, Load Behaviour Notes for Quality of Electrical Supply Continuing Professional Development Course, 2022.)*

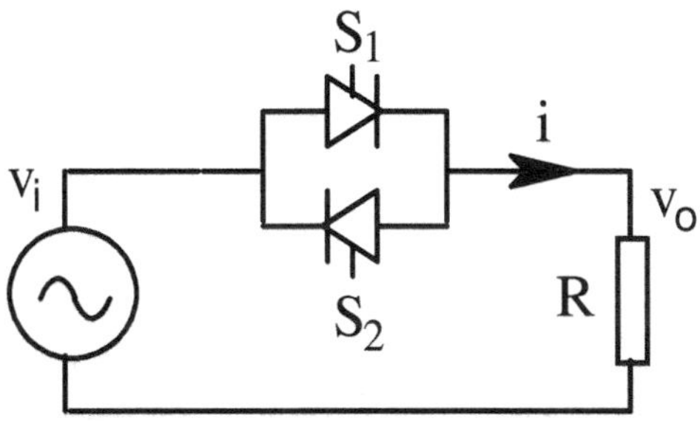

Fig. 7.2 Example of application of an AC static switch for a resistive load. *(From V.J. Gosbell, Load Behaviour Notes for Quality of Electrical Supply Continuing Professional Development Course, 2022.)*

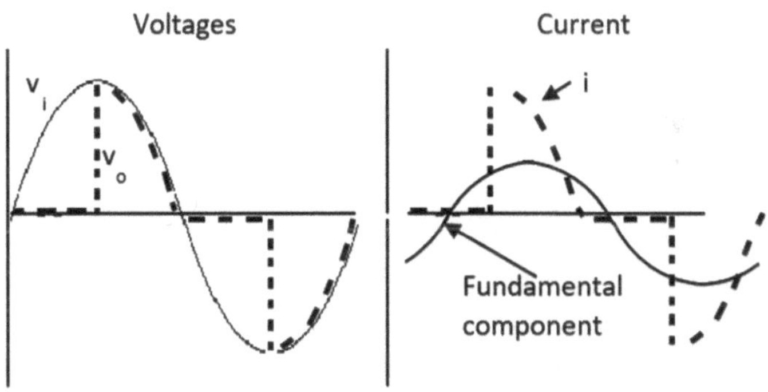

Fig. 7.3 Voltage and current waveforms for AC/AC converter for $\alpha = 90°$. *(From V.J. Gosbell, Load Behaviour Notes for Quality of Electrical Supply Continuing Professional Development Course, 2022.)*

voltage. The moment of firing (i.e. application of the gate pulse) of the SCRs relative to when positive voltage is first applied is called the firing angle, α. By varying α the output voltage can be controlled. As an example, Fig. 7.3 shows the resultant voltage (V_i is input voltage and V_o is output voltage) and current waveforms for the load as shown Fig. 7.2 for $\alpha = 90°$. In this example the effective output voltage will be half the input voltage.

The applications for AC phase control include incandescent lamp dimmers, temperature control of resistive element ovens, some types of static VAr compensators (SVC) and motor soft starters. A schematic for a soft starter application is given in Fig. 7.4. At motor start, the firing angles of the SCRs are adjusted to reduce the RMS voltage applied to the motor. This reduces the input (inrush) current at the expense of a reduction in the available torque. As motor speed gradually increases, the firing angles are reduced to zero (the normal operating value). A consequence of this system is that some current distortion occurs during motor starting.

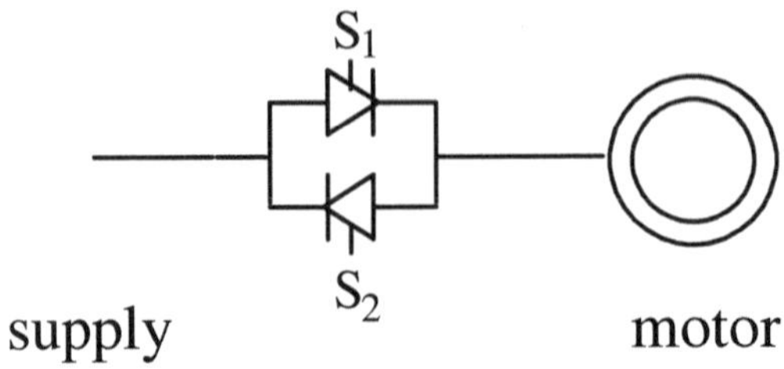

Fig. 7.4 Schematic for AC motor soft starter. *(From V.J. Gosbell, Load Behaviour Notes for Quality of Electrical Supply Continuing Professional Development Course, 2022.)*

7.2.2 AC/DC conversion

AC/DC converters are typically termed rectifiers. These devices typically form the front end of more complex circuits such as switch-mode power supplies (SMPS), compact fluorescent lamps (CFL) and AC motor drive systems (inverters). There are two main rectifier technologies which are characterised by the component which is used to 'smooth' the DC voltage present at the output of the diodes which provide the conversion from AC to DC voltage. These two technologies are capacitor filter (i.e. a capacitor is present at the output of the diodes) and inductive filter (i.e. an inductor is present at the output of the diodes). The nature of the filter component has important implications for the harmonic current emissions of the rectifier circuit.

The DC output voltage V_d of any rectifier can be varied by changing the firing angle α of the rectifier. For example, the DC output voltage for a three-phase full bridge rectifier is as follows:

$$V_d = 1.35x\ V_{LL} \cos\alpha$$

where V_{LL} is the RMS value of the line-neutral voltage applied to the rectifier.

7.2.2.1 Capacitor-filtered rectifier

The basic circuit layout for a single-phase diode bridge rectifier with capacitor filter is shown in Fig. 7.5 whilst Fig. 7.6 shows the circuit symbol for a diode bridge rectifier.

A basic single-phase capacitor-filtered diode bridge rectifier circuit produces the input current waveform and current harmonic spectrum shown in Fig. 7.7. The displacement power factor of diode bridge rectifier circuits is close to unity or may even have a slightly leading displacement power factor. For this single-phase example it can be seen that the 3rd and 5th harmonics form a significant component of the total current. For simple full wave diode

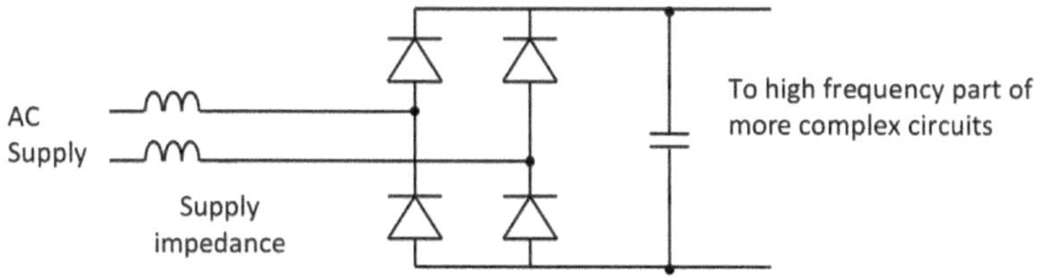

Fig. 7.5 Basic circuit layout for single-phase diode bridge rectifier with capacitor filter. *(From V.J. Gosbell, Load Behaviour Notes for Quality of Electrical Supply Continuing Professional Development Course, 2022.)*

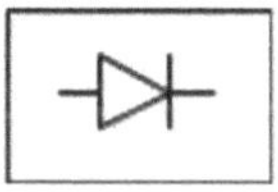

Fig. 7.6 Diode bridge rectifier circuit symbol. *(From V.J. Gosbell, Load Behaviour Notes for Quality of Electrical Supply Continuing Professional Development Course, 2022.)*

bridge rectifiers even order harmonics (2, 4, 6 and so on) are absent in the line current waveform. It should be noted that odd order triplen harmonics (3rd, 9th, 15th and so on) are also absent in the current waveform balanced three-phase rectifiers (when supplied with an undistorted input voltage).

7.2.2.2 Inductor-filtered rectifier

Inductor-filtered rectifiers are used in applications such as battery chargers, the DC supply to smelters and the front end for DC motor drives. The basic circuit layout for a three-phase inductor-filtered SCR rectifier is given in Fig. 7.8 together with its line current waveform.

The current harmonic emissions for a simple three-phase inductive-filtered diode bridge rectifier are given by the following expression:

$$I_n = 0.78\, I_d / n$$

where

- N is the harmonic order = 1,5,7,11,13, and so on.
- I_n is the harmonic current for harmonic order n.
- I_d is the DC current.

Based on the above, Fig. 7.9 shows the harmonic spectrum for a three-phase inductive-filtered rectifier.

The operation of this type of rectifier can also lead to notching of the voltage waveform which can cause maloperation of equipment that rely on the voltage zero crossing for timing synchronisation. Notching occurs as a result of line inductance. As the current from one phase of the rectifier commutates to the next phase, the two phases are momentarily shorted with

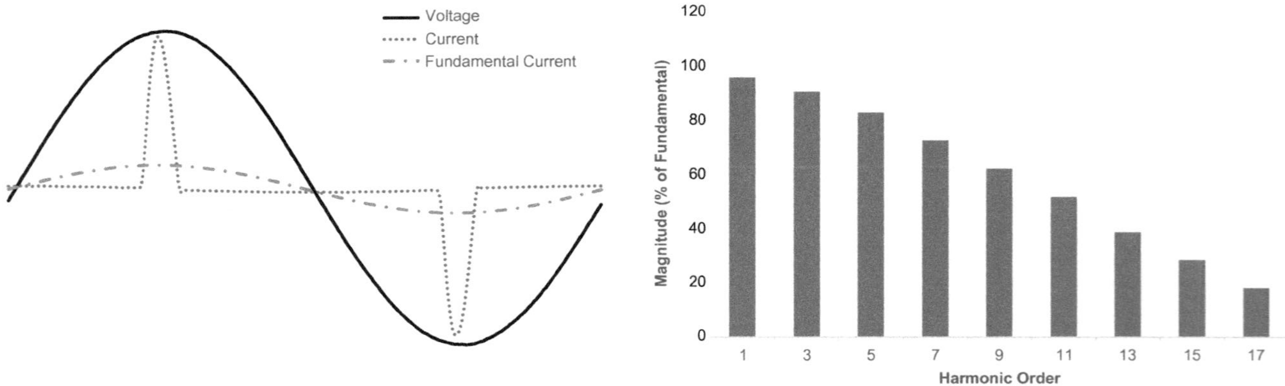

Fig. 7.7 Waveforms and harmonic current spectrum for typical single-phase diode bridge rectifier circuit.

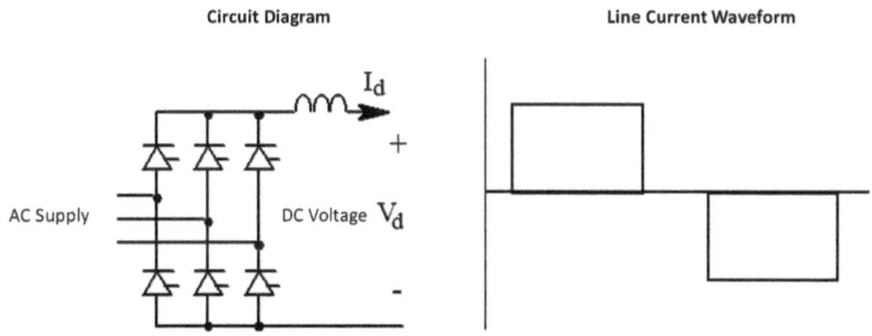

Fig. 7.8 Circuit diagram and line current waveform for three-phase inductor-filtered diode bridge rectifier. *(From V.J. Gosbell, Load Behaviour Notes for Quality of Electrical Supply Continuing Professional Development Course, 2022.)*

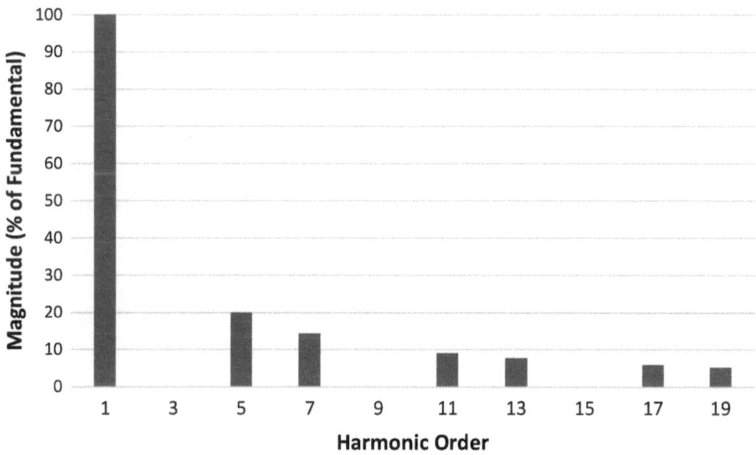

Fig. 7.9 Characteristic harmonic spectrum for inductor-filtered rectifier.

the short-circuit current only being limited by the line inductance. A significant momentary voltage drop results. Fig. 7.10 shows voltage waveform notching due to the operation of a large 6-pulse rectifier.

7.2.3 DC/AC conversion (inverter)

A circuit used to convert DC to AC is called an inverter. This type of converter is used in Compact Fluorescent Lamps (CFLs), AC motor drives and Uninterruptible Power Supply (UPS) systems. Inverters are also widely used in renewable energy generation systems including both wind and solar farms.

A single-phase inverter supplying a resistive/inductive load is shown in Fig. 7.11 together with typical load voltage and current waveforms

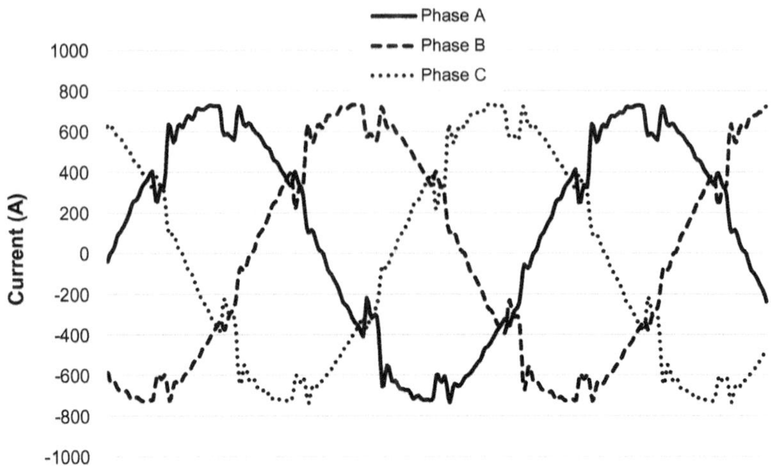

Fig. 7.10 Example of voltage waveform notching caused by operation of a large 6-pulse rectifier.

(V_o and i_o). The inverter uses Insulated Gate Bipolar Transistors (IGBT) appropriately switched at frequencies up to tens of kilohertz to apply a pulse width modulation (PWM) voltage waveform to the load.

Inverters can be controlled to produce variable voltage magnitude and frequency. This makes them an important component of AC motor drives as they can increase the energy efficiency of the load by varying voltage and frequency so that it closely matches the load requirements. Due to their high switching frequencies, inverters produce few low-frequency harmonics but can emit frequencies in the range of 9–100kHz.

7.2.4 More complex conversion circuits

7.2.4.1 Switch-mode power supply

Switch-mode power supplies (SMPS) are used to produce one or more regulated DC output voltages generally in the range 5 to 20V by means of a DC–DC converter in the switching and regulating stage which follows the input rectifier and capacitor. Isolation is usually provided by a high-frequency transformer. A block diagram for a typical SMPS is shown in Fig. 7.12.

SMPS are widely used in a wide range of applications, including computers, consumer electronics and a wide range of industrial applications.

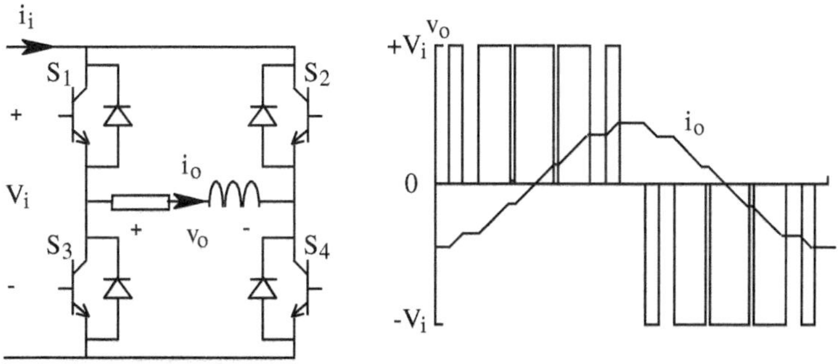

Fig. 7.11 Circuit diagram and voltage and current waveforms for an inverter. *(From V.J. Gosbell, Load Behaviour Notes for Quality of Electrical Supply Continuing Professional Development Course, 2022.)*

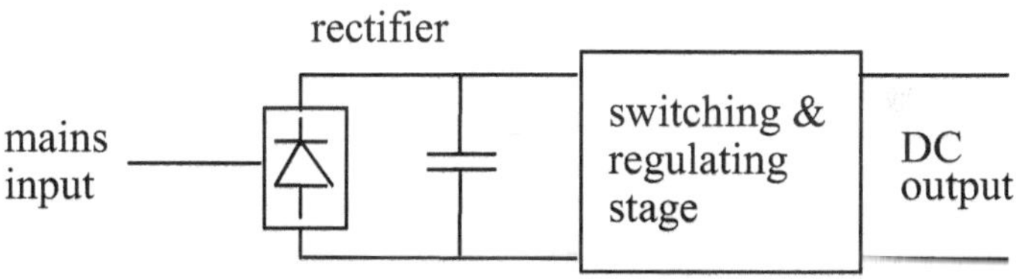

Fig. 7.12 Typical switch-mode power supply schematic. *(From V.J. Gosbell, Load Behaviour Notes for Quality of Electrical Supply Continuing Professional Development Course, 2022.)*

7.2.4.2 AC motor drive

AC motor drives, also known as variable speed drives (VSD) or variable voltage, variable frequency (VVVF) drives, are used to control the speed and torque of induction motors in industrial situations as well as domestic air conditioners and in more recent times in consumer products such as washing machines and refrigerators. These drives approximate the highly controllable characteristics and low speed torque of DC motors whilst providing a low maintenance system. In a motor starting scenario, the drives also mitigate the inrush current transients associated with direct-on-line starting transients. A three-phase AC motor drive typically has two stages:

1. A three-phase rectifier with a capacitive filter.

2. An inverter which provides a variable frequency supply to the motor.

Fig. 7.13 shows the schematic for an AC motor drive.

The characteristics of the current waveform and spectrum of a three-phase AC motor drive are dominated by the rectifier front end and an example current waveform and harmonic spectrum are given in Figs 7.14 and 7.15, respectively. In terms of the current harmonic spectrum it can be seen that there are relatively high emissions for harmonic orders to the limit of

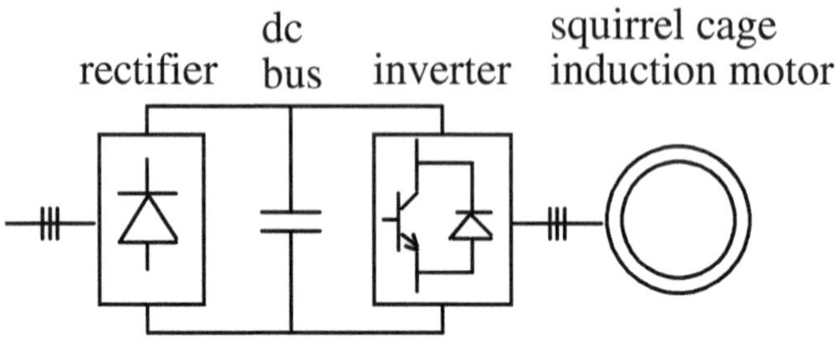

Fig. 7.13 AC motor drive schematic. *(From V.J. Gosbell, Load Behaviour Notes for Quality of Electrical Supply Continuing Professional Development Course, 2022.)*

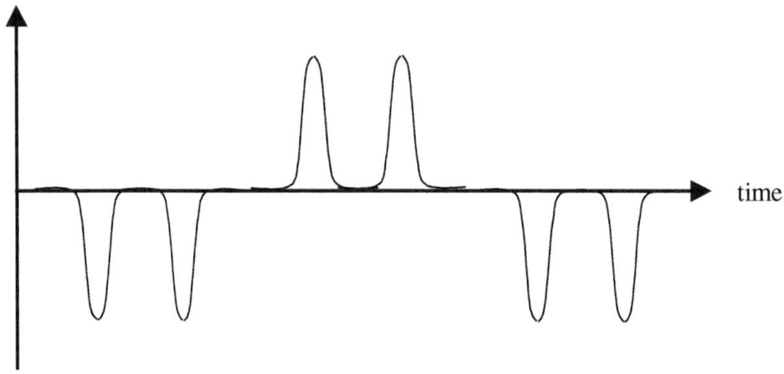

Fig. 7.14 Supply current waveform for three-phase AC motor drive. *(From V.J. Gosbell, Load Behaviour Notes for Quality of Electrical Supply Continuing Professional Development Course, 2022.)*

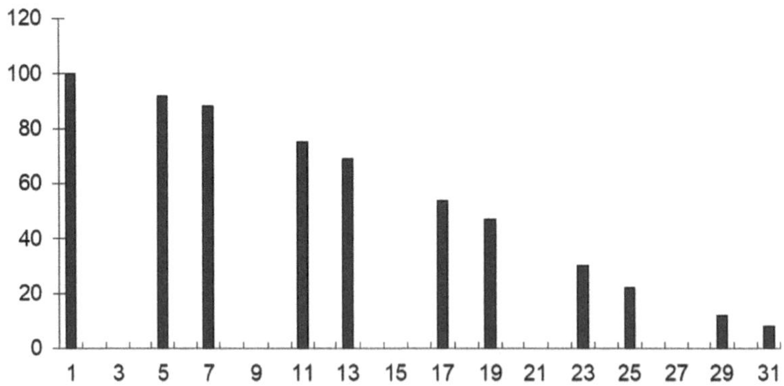

Fig. 7.15 Spectrum of harmonics of the AC side input current waveform of an adjustable speed induction motor drive. *(From V.J. Gosbell, Load Behaviour Notes for Quality of Electrical Supply Continuing Professional Development Course, 2022.)*

those displayed reducing in magnitude as harmonic order increases. For balanced three-phase systems (when supplied by sinusoidal voltage) triplen harmonic emissions are absent as are emissions for even harmonic orders.

7.3 Motor loads

There are two main power quality disturbances which can be attributed to motor loads. These are as follows:

- Voltage fluctuations or voltage sags due to the large inrush currents which are a phenomena of direct-on-line starting of induction motors.
- Harmonic distortion due to saturation of the motor magnetic circuit. This less common phenomenon generally occurs for motors that are poorly constructed for the input voltage and load in order to reduce the cost of the device. Direct-on-line motor starting and harmonic distortion for motor loads are examined in further details as follows.

7.3.1 Direct-on-line motor starting

One of the most significant impacts of motors on power quality is voltage fluctuations or in extreme cases voltage sags caused by direct-on-line starting. During a direct-on-line start, a motor will draw an inrush current which is multiple times the rated current. In some circumstances this inrush current can produce either a voltage fluctuation (rapid voltage change) or a voltage sag. If the depression in the voltage (sag) is severe enough the implications include inability of the motor to start and/or tripping of other nearby loads. If starting occurs frequently enough, the voltage fluctuations may result in flicker. Fig. 7.16 shows an RMS trace of the voltage and current inrush transients for direct-on-line energisation of a relatively small induction motor. This figure clearly shows that the starting current is multiple times the rated current and also demonstrates the dip in the voltage magnitude during starting.

7.3.2 Harmonic distortion

Properly sized and constructed induction motors are generally a low distortion load. As an example, Figs 7.17 and 7.18 show the current waveform and current harmonic spectrum, respectively, for a pedestal fan which is effectively a single-phase induction motor load. It can be seen that the waveform is low in distortion with the current THD being less than 5%. However, if motor saturation occurs, the current waveform will become distorted. As an

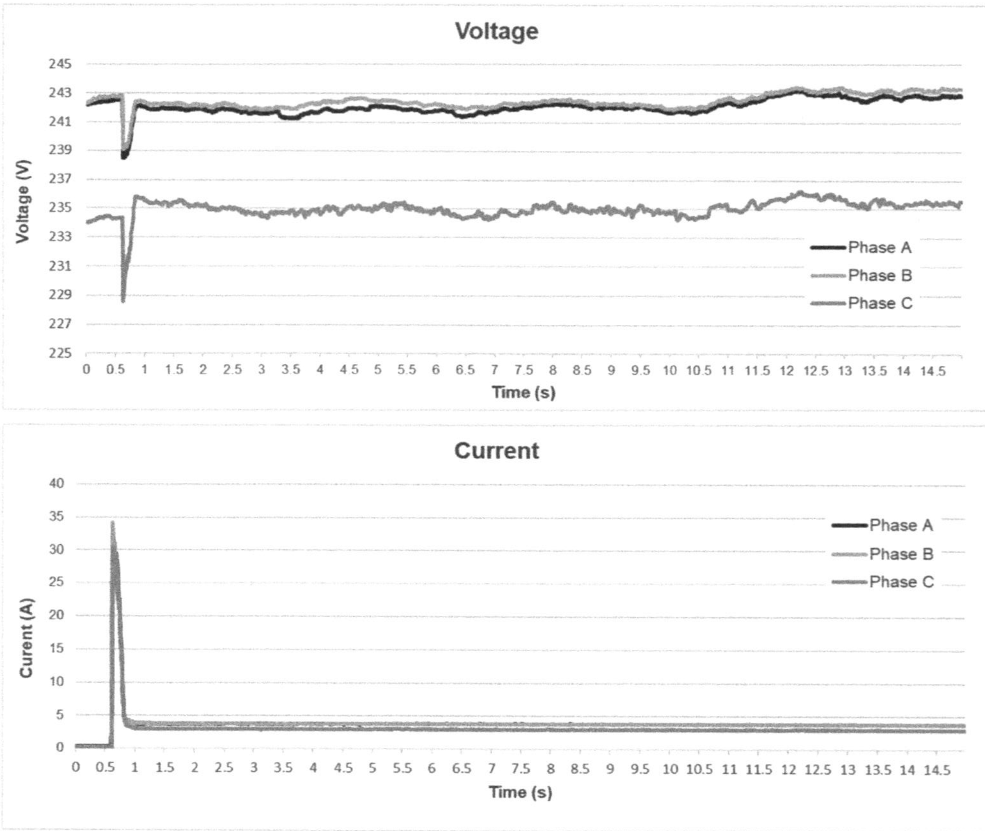

Fig. 7.16 Inrush current transient for direct-on-line motor start (voltage top, line current bottom).

Fig. 7.17 Current waveform for pedestal fan.

example, Figs 7.19 and 7.20 show the current waveform and current harmonic spectrum for a direct-on-line portable air conditioner, respectively. Here it can be seen that the current waveform has some degree of waveform distortion. The current THD for this waveform is 16% which is more than three times the current THD observed for the fan example. Closer examination of the waveform reveals that there is half-wave asymmetry which suggests even order harmonics. This is confirmed by the current harmonic spectra which show that the 2nd harmonic is the dominant order. The characteristics of the waveform and harmonic spectrum indicate distortion due to saturation of the air conditioner motor. This type of behaviour is generally due to reduced amounts of iron in the motor core in order to reduce the manufacturing cost of the appliance.

7.4 Capacitor banks

Capacitor banks are installed by network operators at various locations in their transmission and distribution networks to provide reactive power for voltage support. They are also used for power factor correction in a range of commercial and industrial load settings. The capacitors are switched in and out depending on the level of reactive support required at any given time. There are three main power quality disturbances that can be attributed to the operation of capacitor banks as follows:

- Increases in steady-state voltage
- Switching (oscillatory) transients
- Harmonic resonance

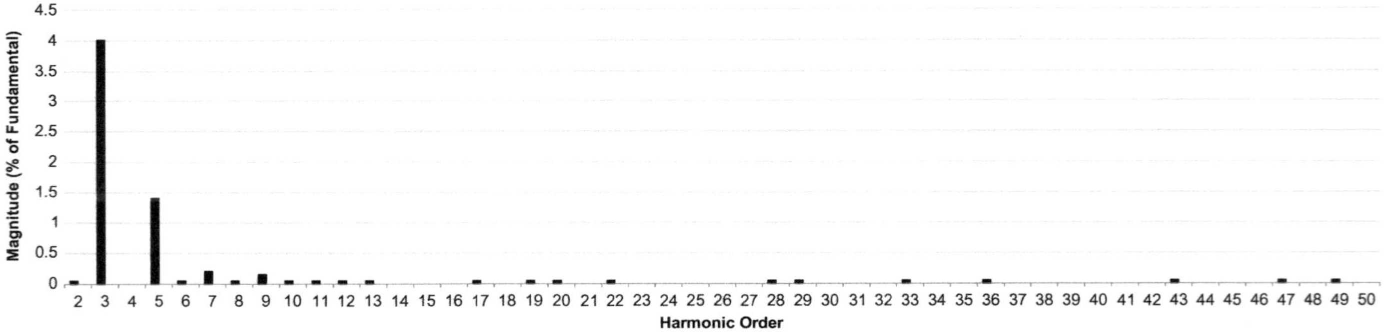

Fig. 7.18 Current harmonic spectrum for pedestal fan.

Fig. 7.19 Current waveform for portable air conditioner.

Issues related to voltage management (i.e. increased voltage) in the presence of capacitor banks are discussed in Chapter 2 and will not be examined further in this chapter. Further detail regarding oscillatory transients and harmonic resonance due to capacitor banks is provided as follows.

7.4.1 Oscillatory transients

An oscillatory transient is a sudden, nonpower frequency change in the steady-state condition of voltage, current or both that includes both positive and negative polarity values. When a discharged capacitor is energised, the voltage of the busbar to which it is attached will momentarily collapse with the depth of the collapse limited only by the reactance in the switched circuit (i.e. when switched, the capacitor effectively approximates a short circuit and a large current flows). This occurs because the voltage across a capacitor cannot change instantaneously (remembering that the voltage–current relationship as a function of time is defined by $i = C\frac{dv}{dt}$) and any short-term change in voltage (in this case energisation) will lead to a very large (theoretically infinite) current. This voltage collapse is followed by an oscillatory recovery that usually lasts for half a cycle or more. The overshoot associated with this oscillation can result in a voltage that has a theoretical peak value of two times the maximum value of the 50-Hz sine wave but typically is no more than 1.5 times. The same effect can occur when a capacitor is switched off and a restrike occurs (i.e. the circuit recovery voltage causes the dielectric between the switch contacts to break down and capacitor current is re-established).

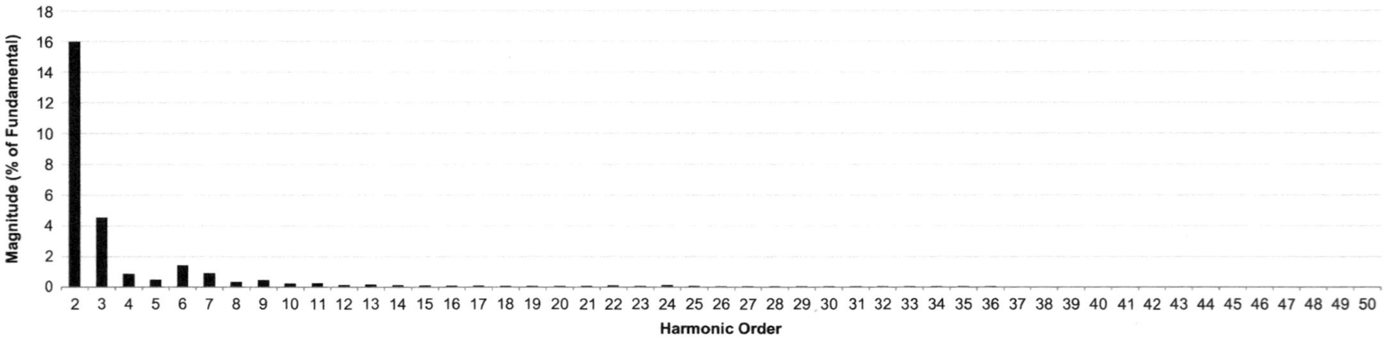

Fig. 7.20 Current harmonic spectrum for portable air conditioner.

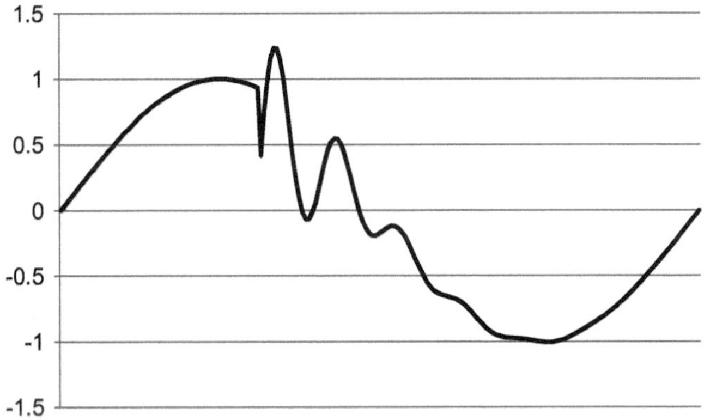

Fig. 7.21 Oscillatory transient typical of capacitor switching.

An example of a capacitor switching waveform is shown in Fig. 7.21. Oscillatory transients consist of a voltage or current whose instantaneous value changes polarity rapidly. Oscillatory transients can be described by their spectral content (predominate frequency), duration and magnitude.

Oscillatory transients with a primary frequency component greater than 500 kHz and a typical duration measured in microseconds (or several cycles of the principal frequency) are considered high-frequency oscillatory transients. A transient with a primary frequency component between 5 and 500 kHz with duration measured in the tens of microseconds (or several cycles of the principal frequency) is termed a medium-frequency transient. A transient with a primary frequency component less than 5 kHz, and a duration from 0.3 to 50 ms, is considered a low-frequency transient.

Normal capacitor energisation in a power system is characterised by a step change in the voltage followed by a transient oscillation in the range 300–1000 Hz. The magnitude of the transient is typically in the range of 1.3–1.8 times the normal peak voltage. Important parameters affecting the characteristics of this transient include

- Capacitor size/location
- Switching device characteristics (resistors, reactors)
- Size/status of other capacitors
- Short-circuit capacity at the substation/at the cap bank
- System load characteristics

Both the initial step change and the oscillation can be important parts of the capacitor switching transient. The initial step change is important because of the high rate of rise. This transient can couple across transformers and into end user facilities by way of the capacitance ratio of step-down transformers, resulting in higher magnitude transients at the end user facilities. Usually, the energy levels in this coupled transient are quite low and conventional surge suppression equipment within the end use facility can generally prevent problems. The oscillation after the initial step change can be more troublesome. The transient magnitude associated with this oscillation is not severe for the transmission or distribution systems due to the fact that equipment and devices on these systems are designed to handle transient voltage magnitudes well in excess of 2.0 per unit. Surge arresters on these systems are designed to protect at around 2.0 per unit and will not be stressed significantly by these transients. However, within end user facilities, the transient oscillations can be a problem, depending on the types of equipment within the facilities. In lower voltage systems, concerns fall into two important categories:

1. Magnification concerns. This is a problem when lower voltage capacitors cause magnification of the energising transient. These lower voltage capacitors could be on the utility distribution system (magnifying the transient for a transmission capacitor bank) or within end user facilities (magnifying the transient for a transmission or distribution capacitor bank).
2. Nuisance tripping of sensitive loads. Even if the transient is not magnified by lower voltage capacitors, some sensitive loads may be affected by the capacitor switching transient. There is enough energy in many of these transients to cause nuisance tripping of adjustable speed drives or other loads with electronic power supplies.

7.4.1.1 Magnification of capacitor energisation transients

Capacitor switching transients can be magnified to quite high values as they pass to a lower voltage level if certain network conditions are met. This can occur when transients originating in the medium voltage (MV) distribution network move into the low voltage (LV) network and there are power factor correction (PFC) capacitors present at LV consumer installations. An example of such a situation is given in Fig. 7.22 which shows an MV/LV network supplied by an MV feeder with inductance L_1. A switched capacitor C_1 is connected to the MV busbar. The series combination of the outgoing MV feeder and the MV/LV transformer has an inductance of L_2. PFC capacitors C_2 are connected to the LV busbar. Magnification occurs when the predominant frequency of the switching transient is approximately equal to the resonant frequency of the LV system.

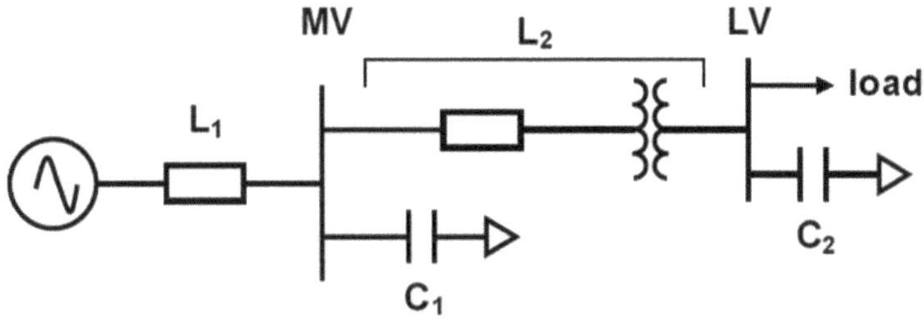

Fig. 7.22 Power system network conducive to magnification of capacitor switching transients.

With reference to Fig. 7.22, this situation can be represented mathematically as follows:

$$L_1C_1 \cong L_2C_2$$

where:

- L_1 = system source inductance seen from MV busbar
- C_1 = capacitance of switched capacitor
- L_2 = inductance of step-down transformer and associated MV feeder
- C_2 = capacitance of LV PFC capacitor

In practice, voltage magnification can occur over a wide range of step-down transformer and PFC capacitor sizes, and can lead to peak voltages on the end user busbar up to 4 times the normal 50-Hz crest voltage.

7.4.1.2 Preventing capacitor energisation transient problems

With an understanding of basic characteristics of capacitor switching transients, it is possible to take steps to prevent or alleviate most capacitor switching problems at the design stage. Some basic design guidelines for both the network operator and the customer can be used to avoid almost all problems.

For the network operator

- The network operator should consider closing control (synchronous closing, pre-insertion resistors or reactors) for banks which are to be energised in proximity to industrial end users who have large numbers of adjustable speed drives or that use low voltage power factor correction.
- The network operator should inform their major industrial end users regarding the location and switching arrangements (e.g. probable operating times and energisation control) of large capacitor banks.

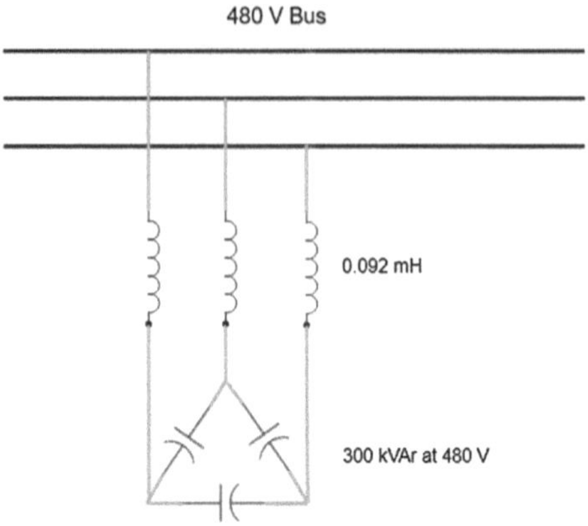

Fig. 7.23 Low voltage capacitor bank installed as harmonic filter.

- The network operator should inform industrial end users that adjustable speed drives should be fitted with line reactors or other means to provide the device immunity from capacitor switching transients.
- The network operator should encourage industrial end users to instal their site power factor correction as a tuned harmonic filter to avoid magnification of capacitor switching transients (Fig. 7.23).

For the customer

- It is the end user's responsibility to protect equipment from the effects of typical disturbances on the supply network.
- Capacitor switching transients on the distribution system can be in the range 1.5–1.8 per unit peak with an oscillation frequency in the range 300–1000 Hz. These transients can occur on a daily basis. End user equipment should not be affected by these transients.
- The equipment manufacturer should be consulted if there are power factor correction capacitors on the same bus. Concerns for both harmonic resonances and magnification of capacitor switching transients should be evaluated.
- Equipment should be installed with sufficient line reactance (on the ac side or on the dc bus) to provide immunity from capacitor switching transients. A 3% input choke is usually sufficient for this purpose as long as there are no low voltage power factor correction capacitors that could magnify the transient.
- The network operator should be contacted before power factor correction capacitors are installed, to determine if there are any large capacitors

banks nearby. The end user should understand that it is possible for its power factor correction capacitors to magnify the existing overvoltage transient caused by a larger capacitor bank within the nearby distribution system. It may be prudent for the end user to install power factor correction with harmonic detuning to prevent this magnification phenomenon (more on this phenomenon in the next section).

7.4.2 Harmonic resonance

A further aspect which needs careful consideration with respect to the use of power factor correction capacitors in a situation where there are significant magnitudes of harmonic currents is the phenomena known as parallel resonance leading to an amplification of the harmonic currents. In the presence of power factor correction capacitors, the total impedance seen by the harmonic currents is the parallel combination of the normal power system impedance (which is linearly proportional to the harmonic order) and the impedance of capacitors (which is inversely proportional to the harmonic order) as shown in Fig. 7.24.

The impedances shown in Fig. 7.24 have the opposite signs and so total impedance increases with the harmonic order quite rapidly to a high value and then falls off as the harmonic order further increases. This is illustrated in Fig. 7.25 for a power supply network for different sizes of power factor correction capacitor banks. Fig. 7.25 also shows the impedance of the same network with no capacitors connected (shown as no capacitors on the graph). It is clear that if there is a load current harmonic which is close to the peaky sections of the curves (i.e. areas of high impedance) a substantial voltage can be developed causing serious voltage distortion.

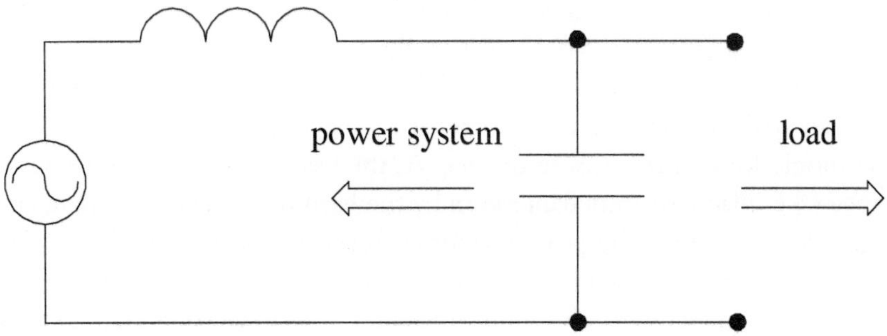

Fig. 7.24 Parallel combination of power system impedance with power factor correction capacitors.

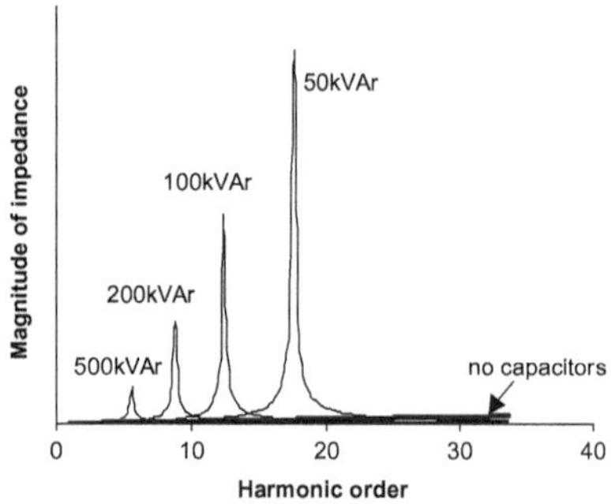

Fig. 7.25 Variation of the impedance with harmonic order for different capacitor banks.

The harmonic order (n) at which the parallel resonance takes place can be calculated using the following expression:

$$n = \sqrt{\frac{S_{FL}}{S_c}}$$

where:

- S_{FL} = fault level at the customer entry point in kVA
- S_c = capacitor kVAr

The earlier equation suggests that the parallel resonance frequency reduces as the capacitor kVAr rating increases and vice versa as shown in Fig. 7.25. For example, at a customer entry point of an industrial installation where the fault level is 2.5 MVA with a 100-kVAr capacitor bank, parallel resonance will take place at 250 Hz (i.e. $n = 5$).

7.4.2.1 Detuning power factor correction capacitors

It is important to avoid parallel resonance at a frequency which is close to a frequency of a harmonic current of a distorting load. The common practice is to detune the capacitor bank so that the lowest order load current harmonic sees a very small impedance. This is achieved by adding an inductor in series with the power factor correction capacitors leading to a situation commonly known as series resonance. At the frequency where series resonance takes place the impedance seen by the harmonic currents is small (see Fig. 7.26) and hence the power factor correction capacitors together with the detuning inductor work as a harmonic filter reducing the voltage distortion. As seen in Fig. 7.26, although parallel resonance still takes place, it happens at a frequency which does not correspond to the lowest frequency of the current harmonic ($n = 5$ in this case) that is present in the load current and hence there will be no serious voltage distortion.

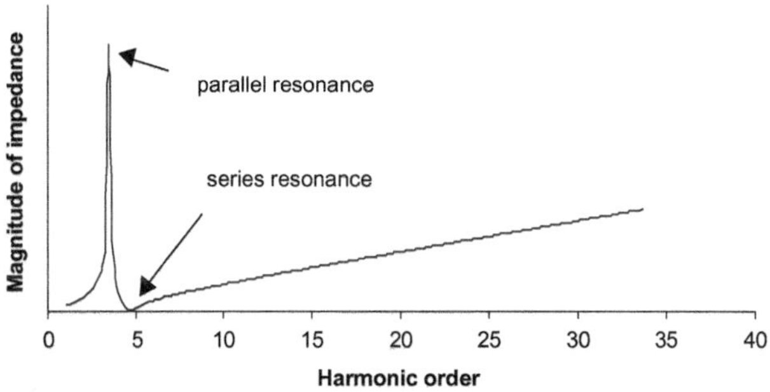

Fig. 7.26 Variation of the impedance with detuning inductor.

It has to be noted that the capacitor still provides the fundamental reactive power whilst working as a harmonic filter. This is quite favourable in a situation where both power factor correction and harmonic filtering are required. However, proper selection of capacitor and detuning inductor is essential for the success of the installation.

Careful system studies covering all configurations of the network and the load must be undertaken before installing power factor correction capacitors even if there are no major harmonic current producing loads in a premise as capacitors can absorb network harmonics thus creating situations not expected in the first place.

7.5 Arcing loads

Arcing loads include arc furnaces, arc welders and discharge-type lighting with magnetic ballasts. Such loads can be a source of voltage flicker and in some cases harmonic distortion.

The main feature of electric arcs that contribute to these effects is their nonlinear voltage–current characteristics. They have a negative resistance for part of their operating cycle (near the current zero point) and are dominated by electrode effects. The current of the arc is limited only by the power system impedance. Section 7.7.6 contains a detailed explanation of the operation of the electric arc furnace.

7.6 Transformers

Transformers can be the source of two main power quality disturbances:

- Voltage sags during energisation.
- Harmonic distortion due to magnetising currents, saturation or DC offset.

Each of the above is examined in further detail as follows.

7.6.1.1 Energisation

Energisation of transformers can cause large inrush currents to flow (6–10 times rated current). Depending on the strength of the local network and the size of the inrush current, this can also lead to voltage sags. The magnitude of the inrush current is related to the remanent flux (i.e. flux remaining when the transformer is de-energised), ϕ_r, within the transformer core and the point on the voltage waveform where the transformer is switched on. Consider the specific case where the transformer winding with N turns is energised using a co-sinusoidal voltage v given by at $t=0$:

$$v = V\cos\omega t$$

Following the energisation, the flux in the core will be given by

$$\phi = \frac{V}{N}\int_0^t \cos\ \omega t\ dt + \phi_r$$

$$\phi = \frac{V}{\omega N}\sin\ \omega t + \phi_r$$

The inrush current is also very high in harmonic distortion (particularly low order even harmonics). Whilst this distortion is only present for a matter of seconds, and as such is unlikely to influence long-term harmonic distortion levels, there is a chance of resonance during the energisation transient leading to high harmonic voltages. As an example, Fig. 7.27 shows three-phase transient current and voltage waveforms for energisation of a 500-MVA transformer. As can be seen, the current waveforms are highly distorted, feature DC offset and do not exhibit the normal 120° phase shift. The voltage sag due to the large inrush current is observable in the voltage waveform. Fig. 7.28 shows an expanded view of the energisation transient showing both the long duration of the transient and the magnitude of the inrush current compared to the magnetising current (i.e. current reached when transient settles). The characteristics of the decaying of the current will be governed by the resistive losses.

7.6.1.2 Harmonic distortion

Whilst in general transformers in steady-state operation are a low emission source for harmonics, there are two circumstances where the harmonic emissions of transformers can be significant:

1. Due to magnetising current.
2. Due to core saturation.

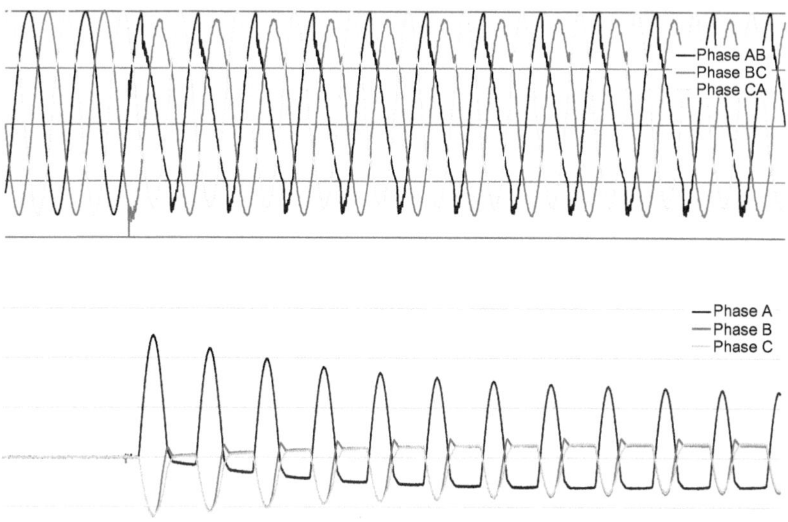

Fig. 7.27 Waveforms for 500-MVA transformer energisation top: current, bottom: voltage.

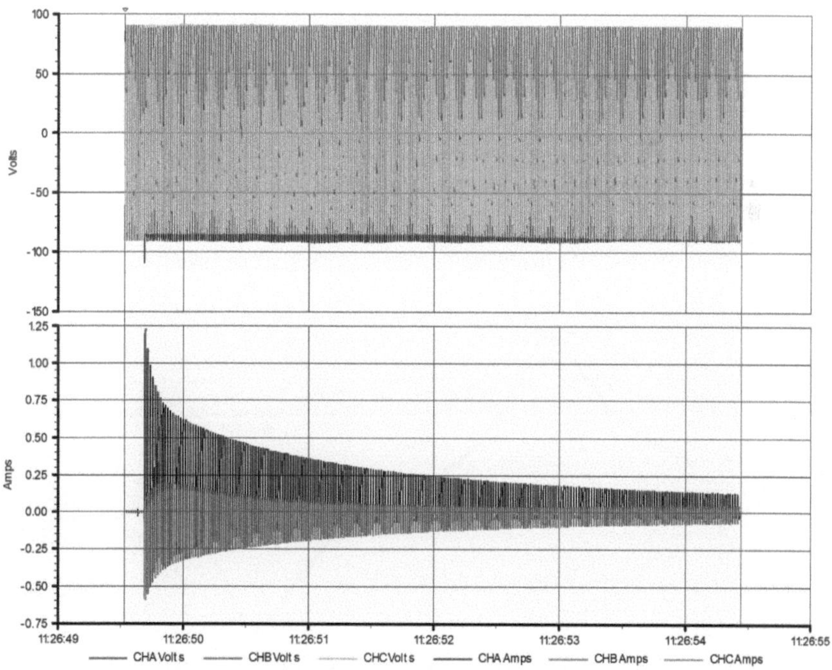

Fig. 7.28 Expanded view of energisation transient for 500-MVA transformer for phase AB voltage and phase A current (top: voltage, bottom: current).

Magnetising current

With respect to magnetising current, harmonics are generated due to the nonlinear magnetising characteristics of the steel core. A typical magnetising current waveform and its spectrum are shown in Figs 7.29 and 7.30, respectively.

The harmonic current associated with magnetisation is not normally considered a problem as magnetising current is typically <1% of rated full-load current. However, harmonic current can become significant when many transformers are operated under lightly loaded, increased voltage conditions. This is discussed in further detail in the next section.

Core saturation

Transformer core saturation can occur due to either applying an excessive voltage to the primary winding of the device, operation at frequencies below the design frequency or exposure to DC offset.

The ferromagnetic core materials used in transformers exhibit a nonlinear flux density (*B*) versus magnetic field strength (*H*) whilst exhibiting a relatively large relative permeability (μ_r) when magnetically unsaturated. However, when saturated this relative permeability reduces substantially. The *B*–*H* relationship of transformer core material is not only nonlinear but also exhibits hysteresis. This nonlinear characteristic will lead to a nonsinusoidal excitation current as the magnetic flux remains sinusoidal resulting from the sinusoidal applied voltage (recall Faraday's law $v = N\frac{d\phi}{dt}$). This means that the larger the applied voltage the greater the peak operating flux density and flux, thus leading to excessive saturation causing a highly distorted and peaky magnetising current. This also leads to significant

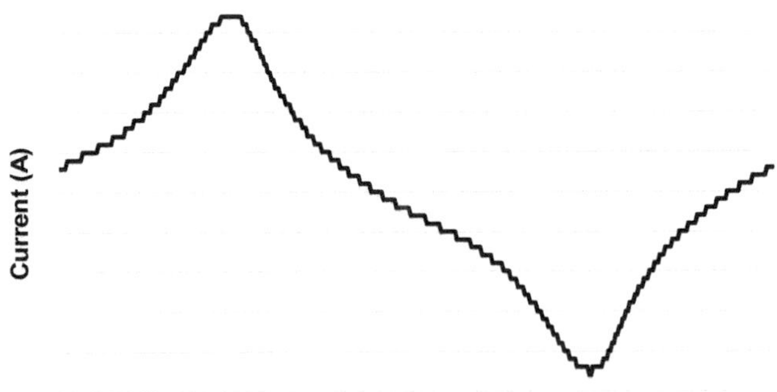

Fig. 7.29 Transformer magnetising current waveform.

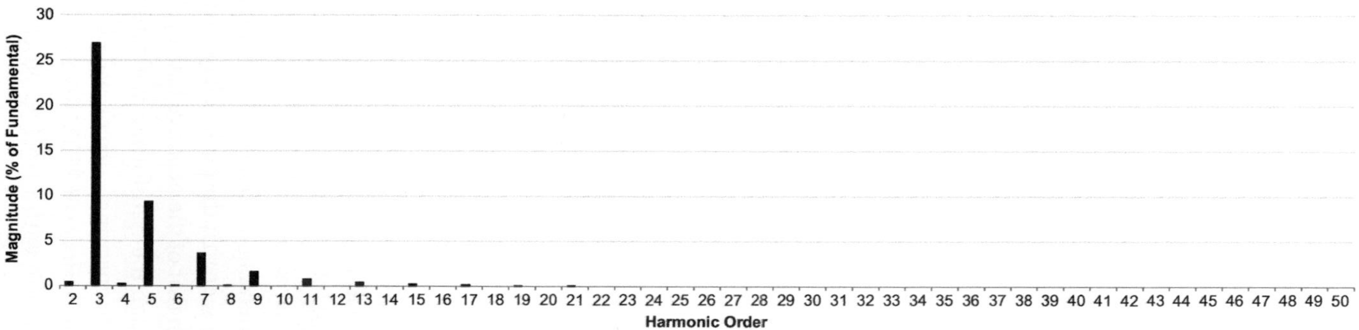

Fig. 7.30 Current harmonic spectrum for transformer magnetising current waveform shown in Fig. 7.29.

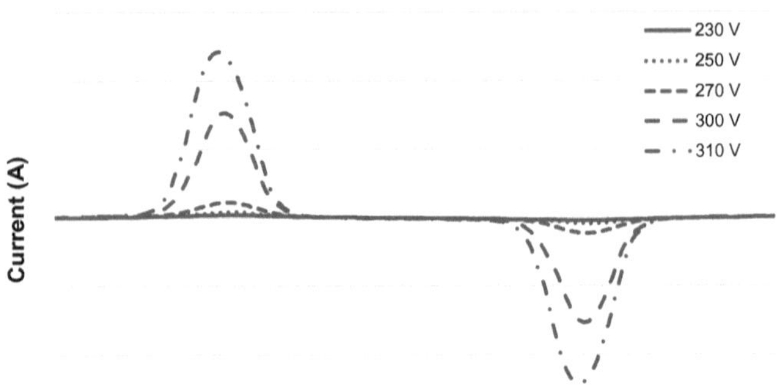

Fig. 7.31 Comparison of magnetising current waveforms for varying primary voltage (rated primary voltage is 230 V).

no-load losses (which includes eddy current and hysteresis losses in the core and copper losses in the excited winding). The transformer design is such that the peak flux density is approximately 1.3 T so that core saturation is limited whilst optimally utilising the core material.

Fig. 7.31 shows the impact of increasing the primary voltage on transformer magnetising current for a transformer with a rated voltage of 230 V. As the transformer core saturates due to the excessive primary voltage it can be seen that both the magnitude and distortion of the magnetising current waveform increase. For this example, the magnetising current exceeds the rated load current when the voltage exceeds 270 V. Fig. 7.32 shows a comparison of the current harmonic spectra for the waveforms shown in Fig. 7.31. The increase in harmonic emission as the transformer core saturates is obvious.

Another cause of transformer saturation is the presence of DC in the primary winding. Any amount of DC voltage dropped across the primary winding of a transformer will cause an additional magnetic flux in the core. This additional flux 'bias' or 'offset' will increase the voltage in one half-cycle and will in turn push the alternating flux waveform closer to the saturation in one half-cycle than the other. Fig. 7.33 shows an example of a waveform with DC offset. It can clearly be seen that the peak voltage in the positive half-cycle is considerably higher than that of the negative half-cycle. If the positive half-cycle voltage is large enough the transformer will saturate.

Fig. 7.34 shows the magnetising current waveforms for a small transformer with a rated primary voltage of 230 V for a sinusoidal waveform with no DC current and a waveform with a 1-V DC offset. The impact of the DC offset can clearly be seen in the shape of the current waveform; the

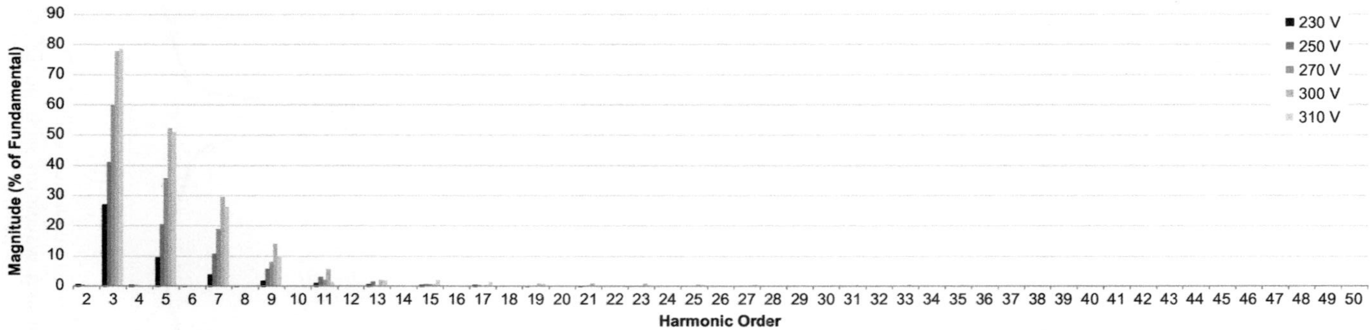

Fig. 7.32 Comparison of current harmonic spectra for waveforms shown in Fig. 7.31.

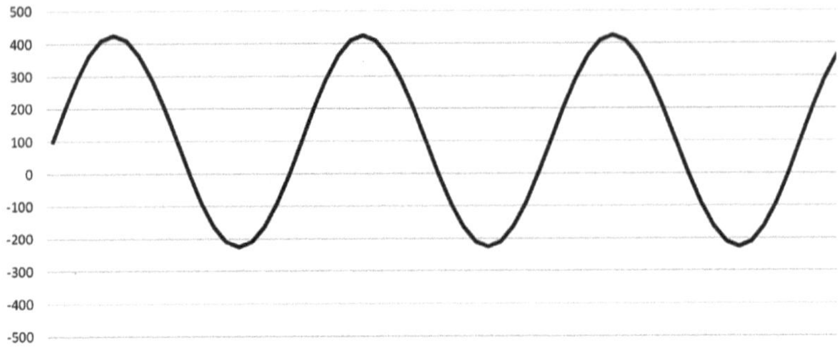

Fig. 7.33 Example voltage waveform containing DC Offset.

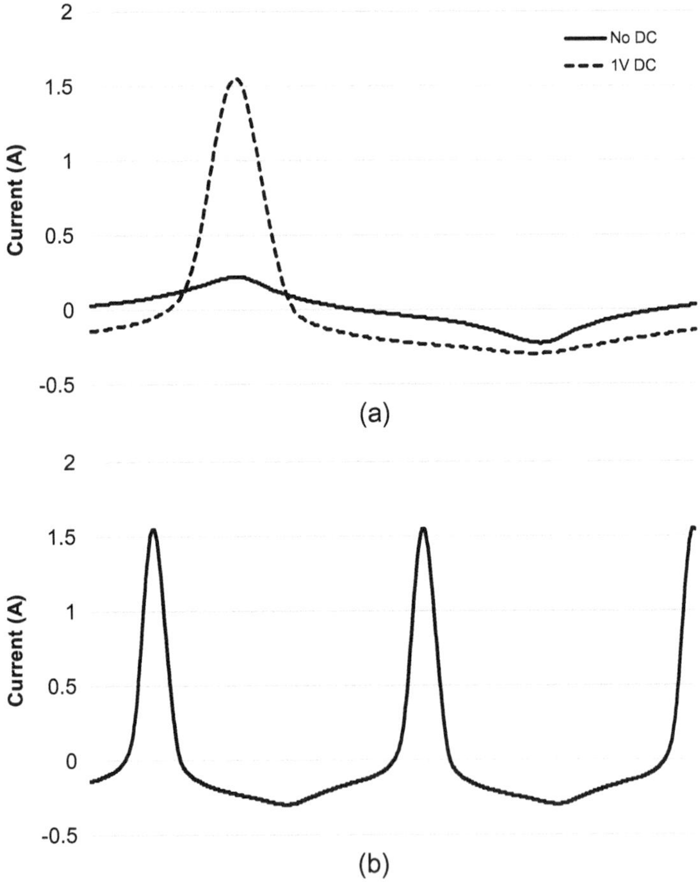

Fig. 7.34 Current waveform for transformer with 1 V DC offset in primary voltage (A) comparison with magnetising waveform with no DC Offset (B) multiple cycles showing impact of DC offset.

magnetising current waveform now lacks half-wave symmetry and the magnitude of the peak current is many times larger. Fig. 7.35 shows the current harmonic spectra for the waveforms shown in Fig. 7.34. Here it can be seen that the DC offset results in significantly higher low order current harmonic components including significant emissions for even order harmonics.

7.7 Load behaviour

This section details the load behaviour with respect to power quality emissions of common equipment connected to electrical power systems.

7.7.1 Lighting systems

7.7.1.1 Incandescent

Incandescent lighting technology can generally be considered as a linear resistive load characterised by a sinusoidal waveform with very low harmonic distortion. Although significant volumes of incandescent lighting technologies are gradually being replaced by more energy efficient and longer lasting modern technologies such as compact fluorescent lamps (CFLs) and more recently light emitting diodes (LEDs), a large amount of incandescent lighting remains connected within electricity distribution networks whether it be in legacy devices or in applications such as heater lamps.

7.7.1.2 Tubular fluorescent

The light output in traditional tubular fluorescent is provided by an arc interacting with mercury gas and a phosphor coating on the outside of the tube. The lamps interface with the network via a ballast which assists with current regulation and in turn stabilisation of the arc. Fig. 7.36 shows the current waveform for a bank of 2×18 W tubular fluorescent lamps. The current THD value for this waveform is 20% and the harmonic spectrum is dominated by the 3rd harmonic as shown in Fig. 7.37.

7.7.1.3 Compact fluorescent lamps

Both CFL and LED lighting technologies interface with the electricity supply by means of a power electronics front end. Whilst supply by power electronics achieves energy efficiencies, this front end is a nonlinear device and as such will draw a distorted current waveform. Laboratory evaluation of CFL performance has indicated that there can be significant variation in the current emission performance across CFLs from different manufacturers [1]. Fig. 7.38 shows photographs of CFL power supply. As can be seen, the power supply is a relatively complex electronic circuit.

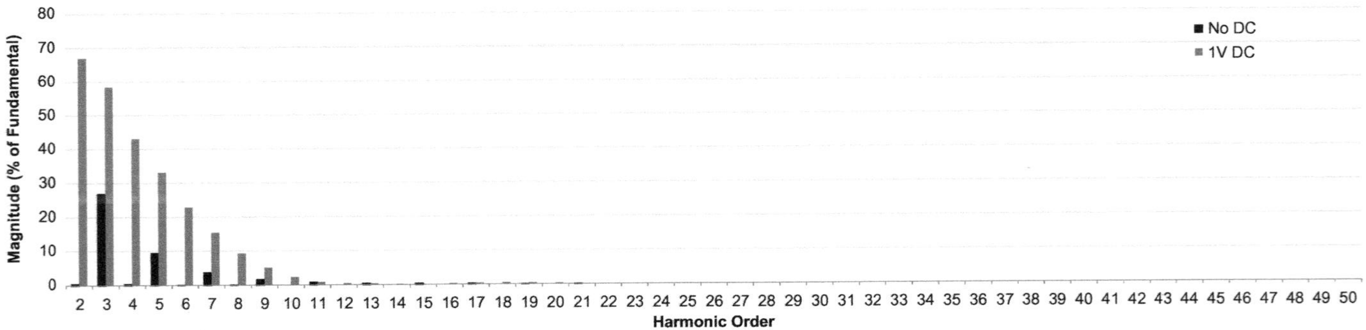

Fig. 7.35 Current harmonic spectra for waveforms shown in Fig. 7.34.

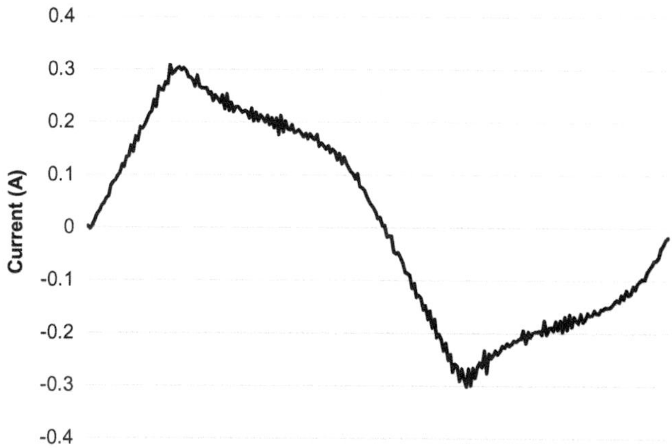

Fig. 7.36 Current waveform for tubular fluorescent lamps.

Fig. 7.39 shows a comparison of the current waveforms for three 14 W CFLs. It can be seen that the current waveforms are highly distorted and the difference in waveform shape can clearly be observed. Fig. 7.40 shows the current harmonic spectrum for the same three CFLs. The current THD for the CFLs is 34%, 154% and 112% for Samples 1, 2 and 3, respectively.

7.7.1.4 LED

Whilst LED lighting is not a new technology it is only in recent times that LED lamps have been deployed in large volumes. LED lamps are significantly more efficient than all other comparable light sources. As an example, consider a 75-W incandescent lamp, the CFL equivalent for this lamp is 19–22 W and the LED equivalent is 7.5 W. Thus there have been two drivers for mass deployment of LED lamps; moves to reduce energy consumption and the decreasing price of LED technology. As with many high efficiency devices, LED lamps interface with the network through means of a complex power electronic converter (example photograph of LED power supply shown in Fig. 7.41). Because of this power electronic converter, LED lighting is a source of harmonic emissions. Fig. 7.42 shows the current waveforms for a selection of LED lamps. It can be seen that the current waveforms are highly distorted and share some similarities with the CFL waveforms shown in Fig. 7.39. The current THD values for the waveforms shown in Fig. 7.42 are as follows: 7.5-W LED: 79%, 10-W LED: 118%, 6.5-W LED: 132%. Fig. 7.43 shows the current harmonic spectrum for each of the LED waveforms shown in Fig. 7.42. It can be seen that there are harmonic emissions for all odd harmonic orders out to the limit of measurement (50th order).

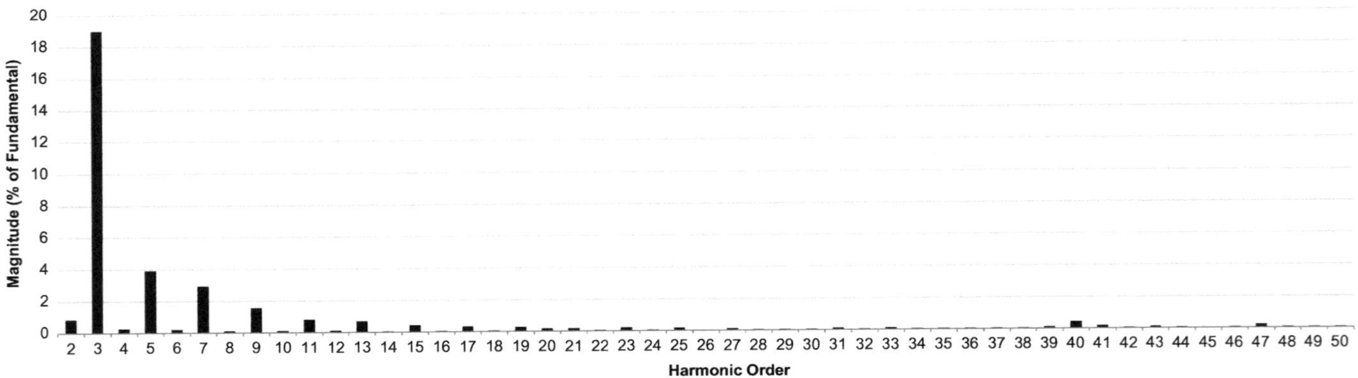

Fig. 7.37 Current harmonic spectrum for tubular fluorescent lamps.

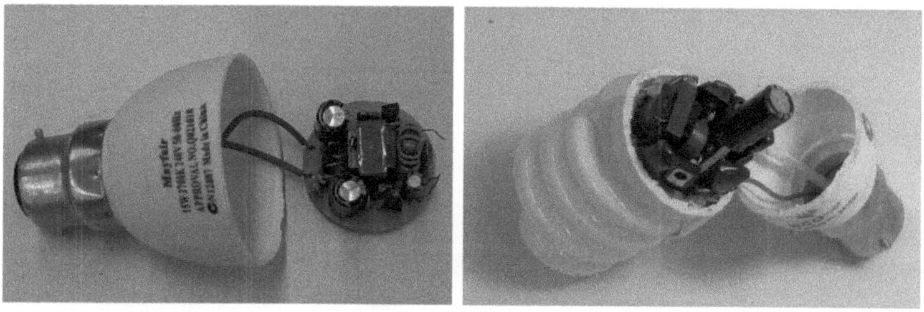

Fig. 7.38 Photograph of CFL power supplies.

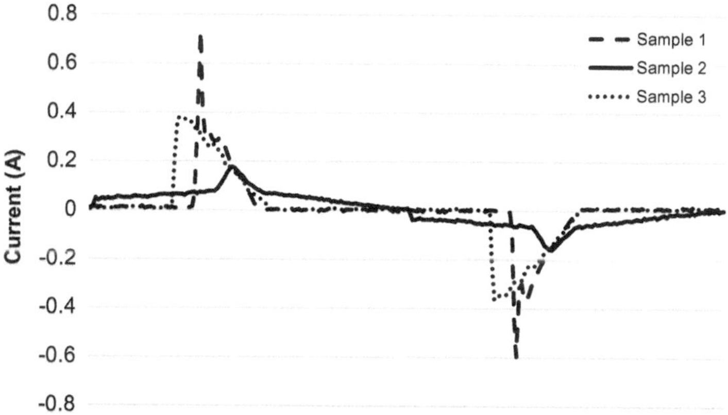

Fig. 7.39 Comparison of CFL current waveforms.

7.7.1.5 Discharge lighting

Discharge lighting is used in applications such as street lighting, warehouse lighting and sporting field lighting. Traditional discharge lighting works by sending an electrical discharge (arc) through an ionised gas to create a plasma which is the source of the light. The ballast used in discharge lighting limits the current to within the capabilities of the lamp and helps to stabilise the arc. Discharge lighting is a known source of harmonic distortion and it is the arc that is the main source, not the ballast. Fig. 7.44 shows the input current waveforms for two street lights which are typical applications for discharge lighting. It can be seen that there is significant waveform distortion particularly for the 150-W metal halide example. The current THD is 9% for the mercury vapour lamp and 15% for the metal halide lamp. There is also evidence of half-wave asymmetry in the metal halide waveform indicating the presence of even order harmonics. Fig. 7.45 shows the current harmonic spectra for the lamps. It can be seen that the spectra are dominated by

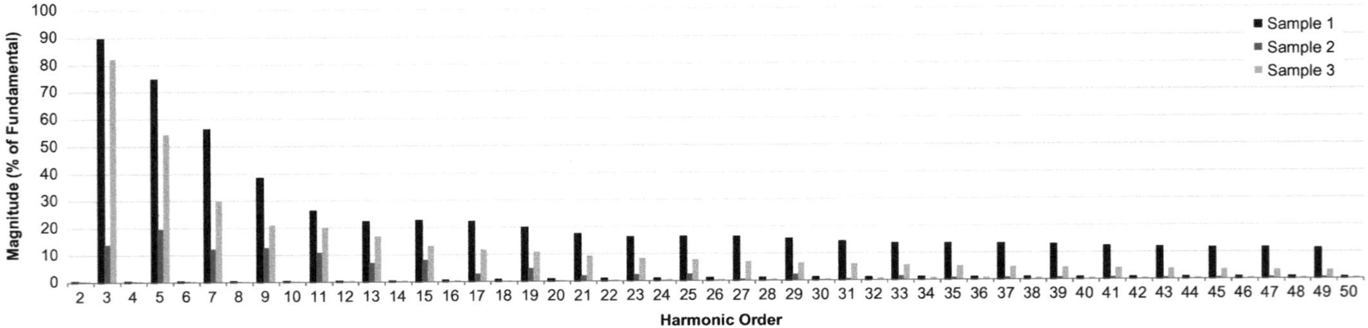

Fig. 7.40 Comparison of CFL current harmonic spectra.

Fig. 7.41 Photograph of LED power supplies.

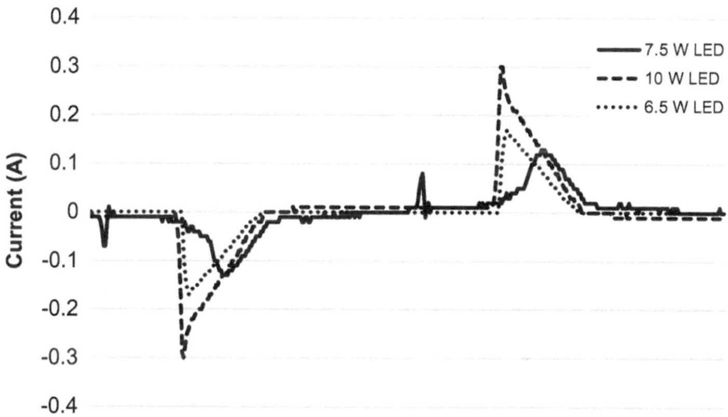

Fig. 7.42 Current waveforms for a selection of LED lamps.

low order harmonics, particularly the 3rd. For the metal halide lamp, the presence of even order harmonics is confirmed with comparatively high emissions for 4th, 6th and 8th harmonic.

7.7.2 Consumer electronics

For the purposes of this chapter, consumer electronics have been categorised into two subcategories as follows:

- Audio-visual equipment—including televisions, DVD players and gaming consoles.
- Information technology equipment—personal computers, computer monitors and printers.

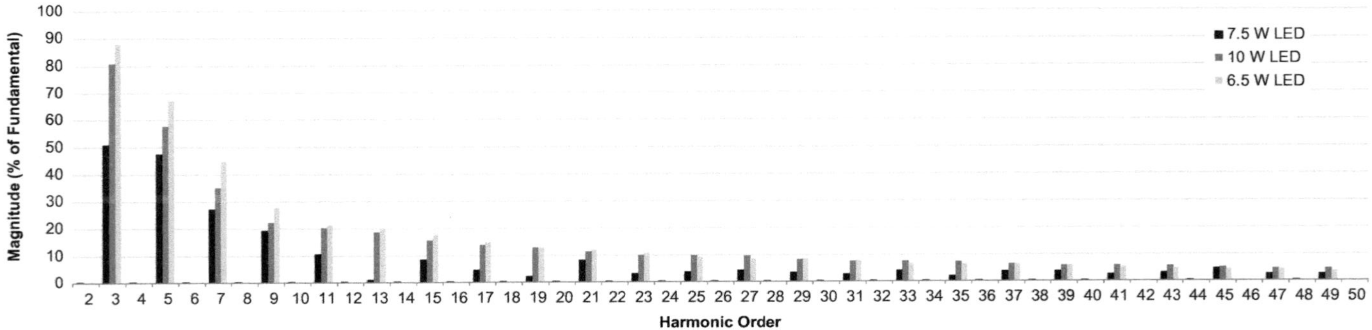

Fig. 7.43 Current harmonic spectrum for LED lamp waveforms shown in Fig. 7.42.

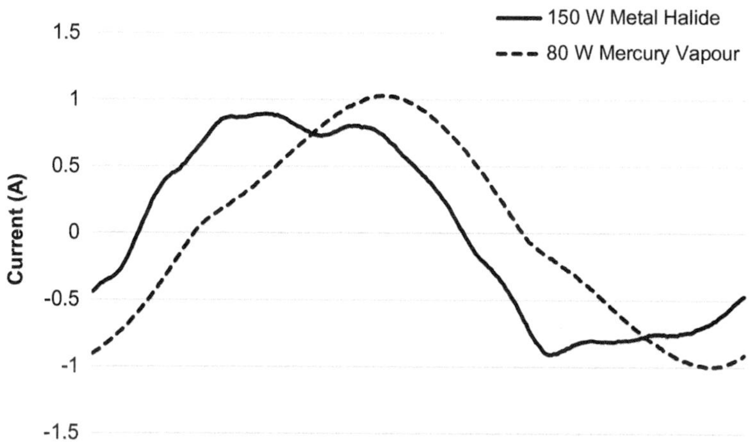

Fig. 7.44 Current waveform for two discharge lighting technologies.

All modern consumer electronics will be supplied using some form of switch-mode power supply and as such will draw current waveforms with some degree of distortion.

7.7.2.1 Audio-visual equipment

Televisions

Television technology has varied considerably over the past 20 years. Cathode ray tube (CRT) technology has been superseded by plasma technology which has subsequently been replaced by LCD technology. Whilst in general, the electricity consumption of televisions has increased, to some extent due to significant increases in screen sizes, there is evidence to suggest that current harmonic distortion emissions have reduced. Figs 7.46 and 7.47 show comparisons of harmonic current waveforms and spectrums for CRT, plasma and LCD televisions. It can be seen that the LCD television has the best performance in terms of waveform distortion; this is reflected in the current THD values for each television technology which are as follows:

- CRT: 94%
- Plasma: 34%
- LCD: 13%

DVD player

Figs 7.48 and 7.49 show the typical current waveform and current harmonic spectrum for a DVD player. As can be seen, the characteristics of this device are that of a simple diode bridge rectifier circuit.

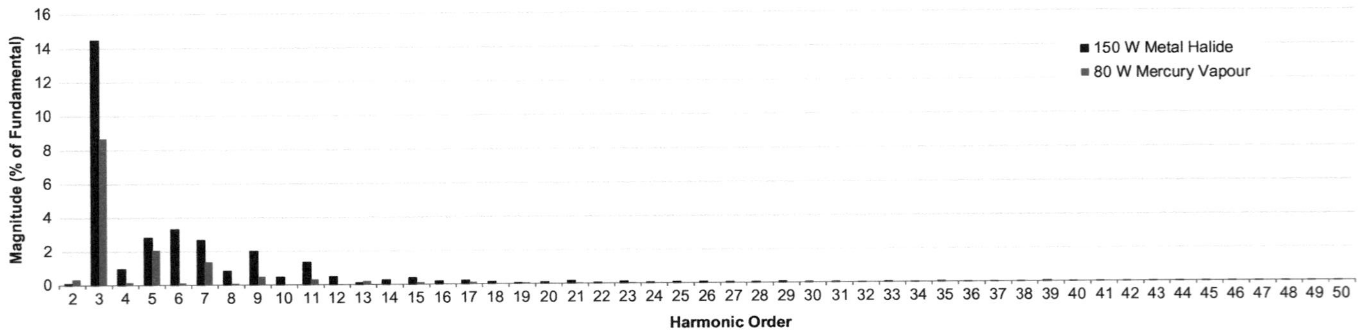

Fig. 7.45 Current harmonic spectra for two discharge lighting technologies.

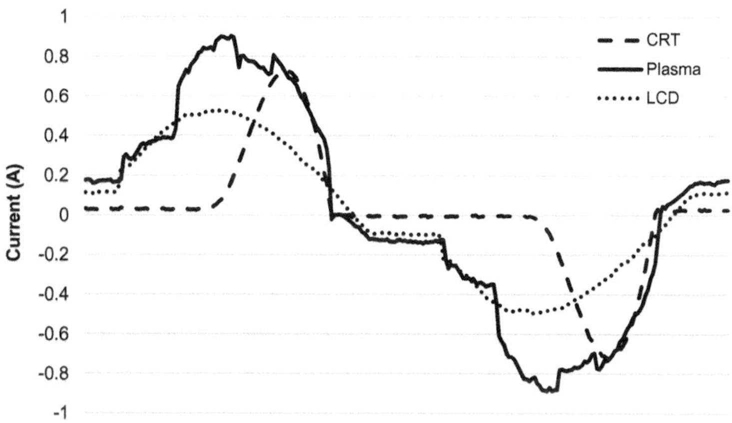

Fig. 7.46 Comparison of television current waveforms.

Gaming console

Fig. 7.50 shows the current harmonic waveform for a common gaming console. It can be seen that this device has a somewhat unusual waveform. As shown in Fig. 7.51 relatively high current harmonic emissions are observed for all odd order harmonics displayed.

7.7.2.2 Information technology equipment

Personal computers

The current waveform and current harmonic spectrum for a desktop and laptop personal computer are shown in Figs 7.52 and 7.53, respectively. It can be seen that both devices have similar waveforms (although the laptop computer draws less current). The waveforms and harmonic spectrum show large amounts of distortion and are characteristic of a simple single-phase diode bridge rectifier circuit. The current THD for the desktop computer is 191% whilst it is 282% for the laptop.

LCD monitor

Fig. 7.54 shows the current waveform for a LCD computer monitor. It can be seen that this device has a current waveform which is typical of an uncompensated single-phase diode bridge rectifier. The current THD for this device is 198% and the device draws significant magnitudes of low order harmonics as shown in Fig. 7.55.

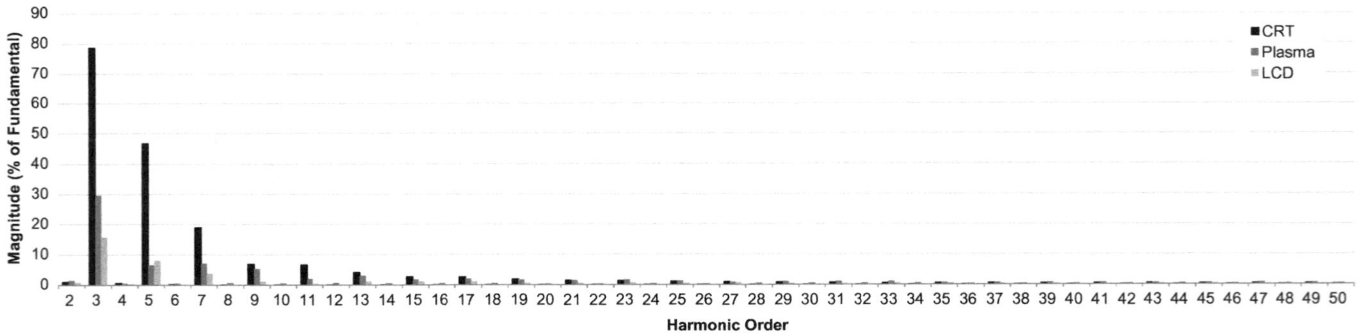

Fig. 7.47 Comparison of television current harmonic spectrum.

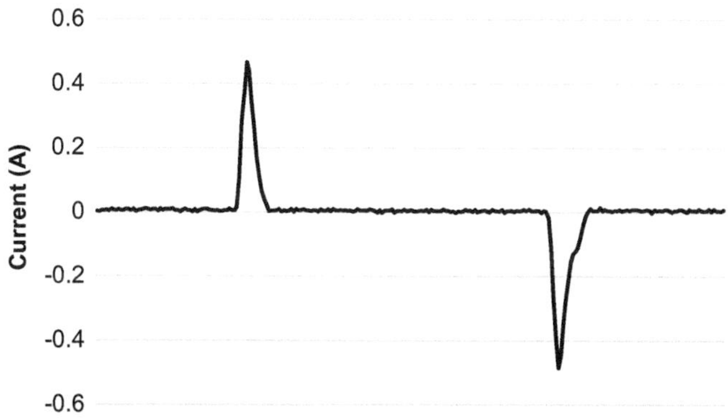

Fig. 7.48 Current waveform for DVD player.

7.7.2.3 Chargers

Small power battery chargers for devices such as mobile phones, tablets and power tools are ubiquitous. All of these devices are examples of AC/DC converters. Fig. 7.56 shows example current waveforms for two different chargers. Sample 1 is a battery charger for an 18 V power tool battery whilst Sample 2 is a charger for a mobile phone. Both current waveforms, but especially that for Sample 2, have high distortion out to high order harmonics as shown in Fig. 7.57. The current THD for Sample 1 and Sample 2 is 160% and 314%, respectively.

7.7.2.4 Microwave ovens

Microwave ovens are a relatively high power electronic device (of the order of 1 kW) found in most residential homes. Fig. 7.58 shows the current waveforms for two sample microwave ovens. It can be seen that the current waveforms are highly distorted with THD values of 31% and 40% for Sample 1 and Sample 2, respectively. Fig. 7.59 shows the current harmonic spectra for these microwave ovens. It can be seen that the spectra are dominated by low order harmonic emissions. The half-wave asymmetry observed for Sample 1 is reflected in the relatively high 2nd harmonic emission for this device.

7.7.3 Air conditioning

Air conditioner technology consists of two main varieties: direct-on-line motor and inverter. Whilst almost all fixed installation split system type domestic air conditioners are now supplied by inverters there remains a significant market for all-in-one and portable air conditioners. These all-in-one air conditioners tend to utilise direct-on-line motors.

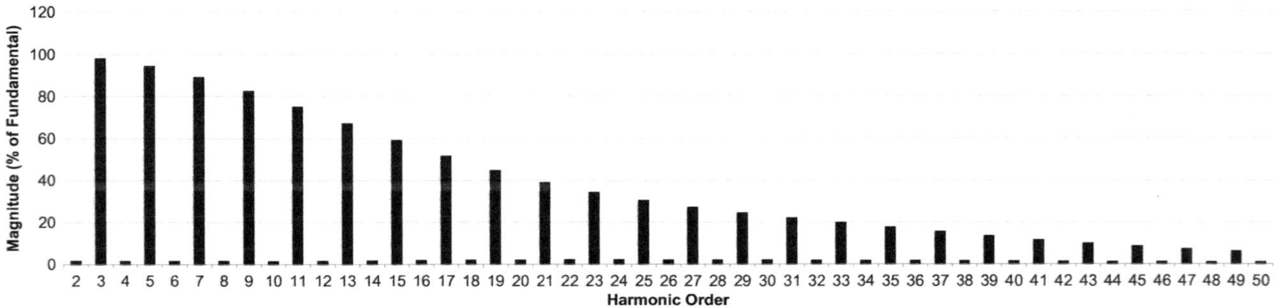

Fig. 7.49 Current harmonic spectrum for DVD player.

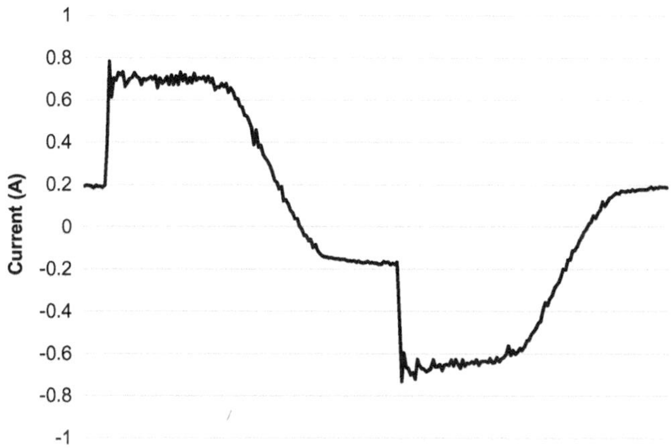

Fig. 7.50 Current waveform for gaming console.

7.7.3.1 Direct-on-line air conditioner

An example of the current waveform and current harmonic spectrum for a direct-on-line air conditioner is shown in Figs 7.19 and 7.20, respectively. As noted, in this example, the induction motor is saturated leading to additional harmonic distortion. Air conditioners with direct-on-line motors will also exhibit the inrush currents which are characteristics of direct-on-line motor starting. As such, they can be a source of voltage fluctuations and/or voltage sags. Fig. 7.60 shows the RMS trace of the inrush current transient for a direct-on-line air conditioner motor start. In this case, the inrush current is approximately five times the rated current.

7.7.3.2 Inverter air conditioner

Fig. 7.61 shows the current waveform for an inverter type split system air conditioner. Whilst first generation inverter air conditioners were somewhat notorious for their harmonic emissions, it can be seen that this waveform is almost sinusoidal with a current THD of only 3%. Fig. 7.62 shows the current harmonic spectrum.

7.7.4 White goods

White goods consist of appliances such as washing machines, clothes dryers and refrigerators. White goods generally contain motors. Traditionally, these motors have been direct-on-line start; however, the trend is now towards the use of variable speed drives in order to achieve increased energy efficiency. Fig. 7.63 shows example current waveforms for refrigerators with

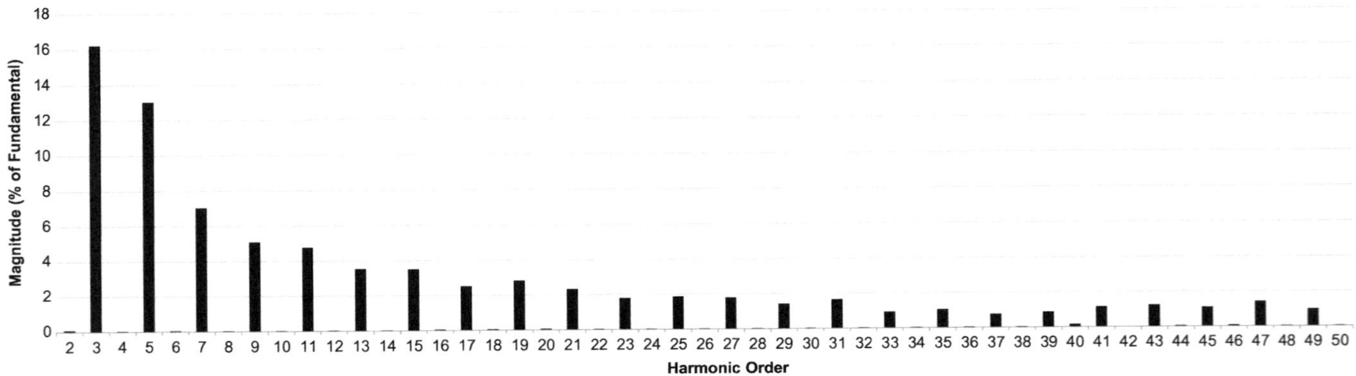

Fig. 7.51 Current harmonic spectrum for gaming console.

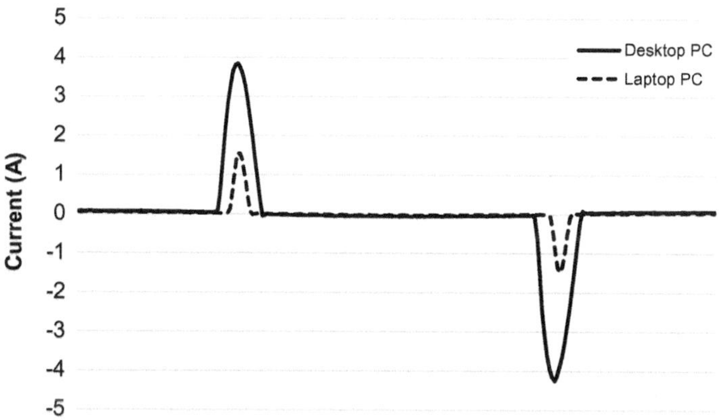

Fig. 7.52 Current waveforms for desktop and laptop personal computers.

direct-on-line and inverter front ends. Fig. 7.64 shows the current harmonic spectra. Whilst inverter technology unquestionably increases energy efficiency, it can be seen that the refrigerator with the inverter front end has a waveform which is much more distorted and has proportionally higher current harmonic emissions for odd order harmonics out to at least the 17th order.

7.7.5 Renewable energy

7.7.5.1 Small-scale renewable energy equipment

There are three main power quality implications that can be attributed to small-scale renewable energy generation, which is dominated by solar photovoltaics. The power quality implications for other distributed energy resources such as electric vehicles and battery energy storage are similar. These three main power quality implications are as follows:

- Voltage rise—this occurs due to export of current into the electricity distribution network. This phenomenon is dealt with in detail in Chapter 2 and will not be examined further here.
- Unbalance—as most small-scale renewable energy generation is deployed as single-phase installations, it is conceivable that they may be a source of network unbalance. However, given the random nature of deployments it is unlikely that installations will be concentrated in any single phase and provided that the underlying loads are evenly distributed (i.e. balanced) and the network impedance is also balanced, it is unlikely that, on average, renewable energy installations will contribute to unbalance.

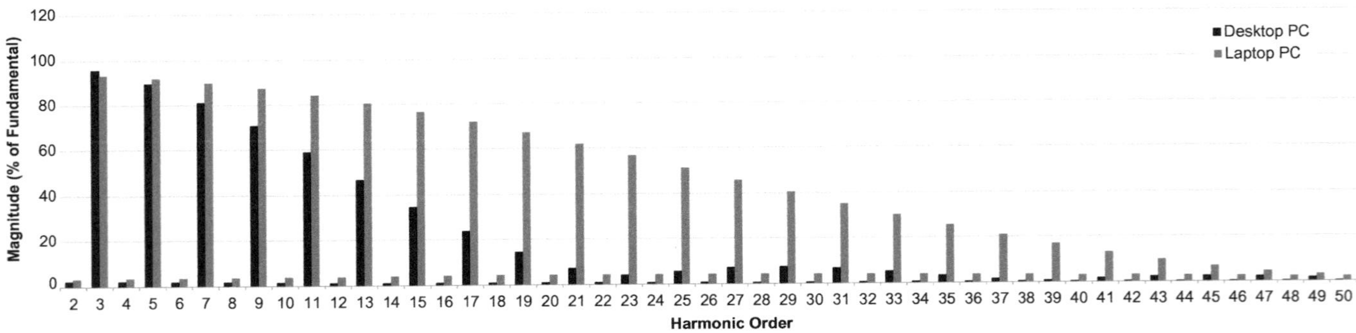

Fig. 7.53 Harmonic current spectrum for desktop and laptop personal computers.

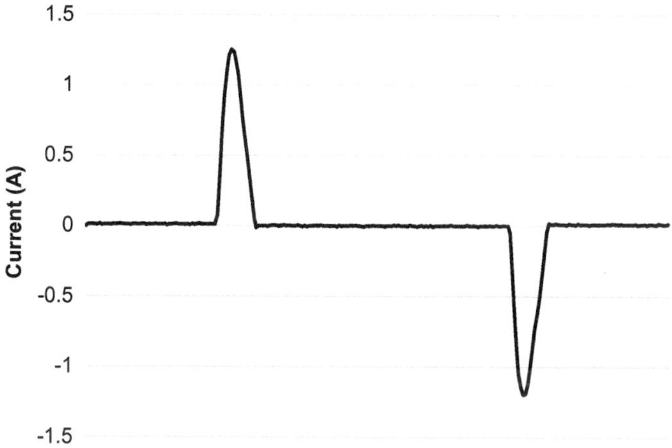

Fig. 7.54 Current waveform for LCD computer monitor.

- Harmonic distortion—the interface between small-scale renewable energy sources (and electric vehicles and battery energy storage) is inverters. Inverters being a power electronic device are a source of harmonic distortion. Further analysis of the harmonic emissions from small-scale renewable energy sources is provided as follows.

Fig. 7.65 shows a photograph of 20-kW solar PV inverters.

Small-scale renewable energy harmonic distortion

Whilst the inverters used to interface renewable energy sources with the electricity distribution network are a power electronic load and as such will be a source of harmonic distortion, in most jurisdictions there are strict standards for the current harmonic emissions from grid-connecting inverters. In order to demonstrate the relatively low levels of emissions allowed for small-scale renewable energy inverters, two comparisons are made between the harmonic emissions allowed for inverters as specified in IEEE 1547 and the current harmonic emissions allowed by IEC 61000-3-2 for other household appliances. The first comparison, as shown in Fig. 7.66, is with equipment classified as Class C (lighting equipment) in IEC 61000-3-2. It could be argued that this comparison is not highly relevant as inverters are a significantly higher powered load than most lighting and likely to be operating for longer periods. To address this argument, the second comparison, as

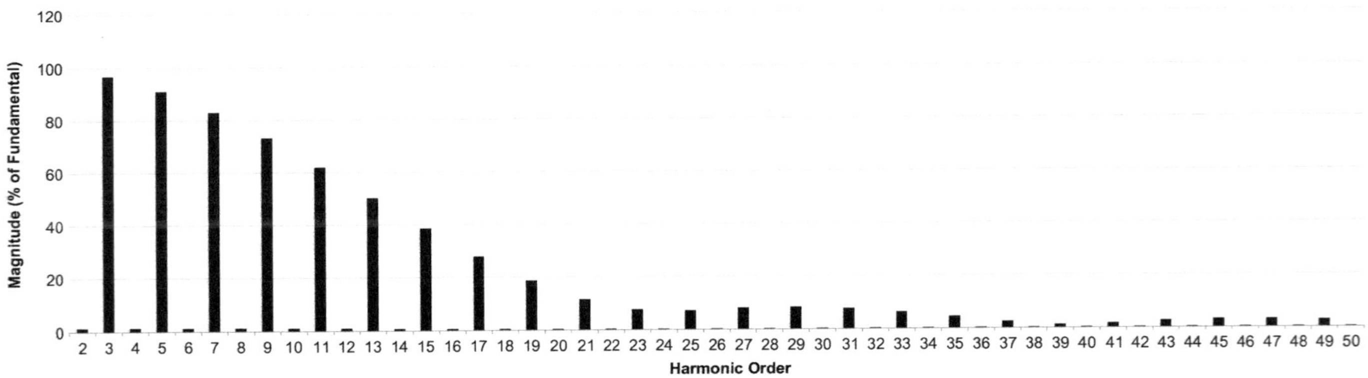

Fig. 7.55 LCD computer monitor current harmonic spectrum.

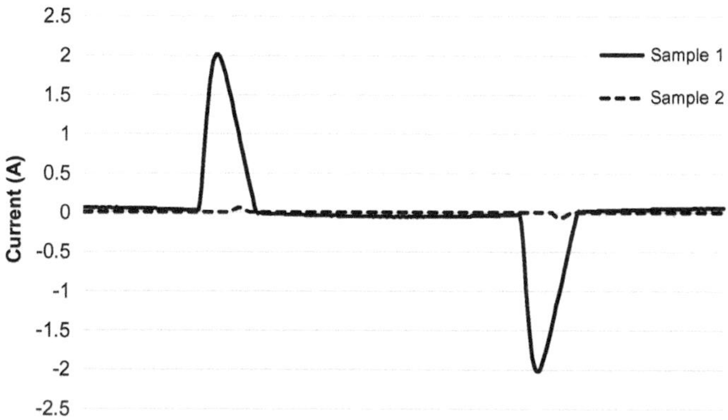

Fig. 7.56 Example current waveforms for chargers.

shown in Fig. 7.67, is with equipment with a rating of 2.5 kW classified as Class B (portable tools and arc welders) in IEC 61000-3-2. This comparison is more in line with the rating of inverters. Regardless of the comparison, it can be seen that the emissions allowed for inverters are smaller, and in some cases much smaller, than those allowed for other household appliances. These relatively strict harmonic current emission limits are likely the reason that at the time of writing there is little evidence of inverters associated with small-scale renewable energy generation causing any issues with respect to harmonic distortion in electricity distribution networks.

Fig. 7.68 shows the current waveforms for a 2.5-kW grid-tie solar photovoltaic inverter from a very well-known manufacturer. Waveforms are shown for two operating points: 1.1 kW (approximately half rating) and 2.2 kW (close to rating). It can be seen that both waveforms exhibit low harmonic distortion with the waveform for the 2.2-kW operating point being almost sinusoidal. Fig. 7.69 shows the current harmonic spectra for the waveforms shown in Fig. 7.68. Emissions for the 3rd order dominate; however, there are emissions out to high orders. The current THD is 4.62% for the 1.1-kW operating point and 3.38% for the 2.2-kW operating point (Fig. 7.68).

Large-scale renewable energy equipment

The inverters used in large-scale renewable energy farms bear little resemblance to the inverters utilised in small-scale applications. These large inverters have much higher power ratings with 2.5 MW being typical for solar farm inverters and wind turbine inverters being of the order of

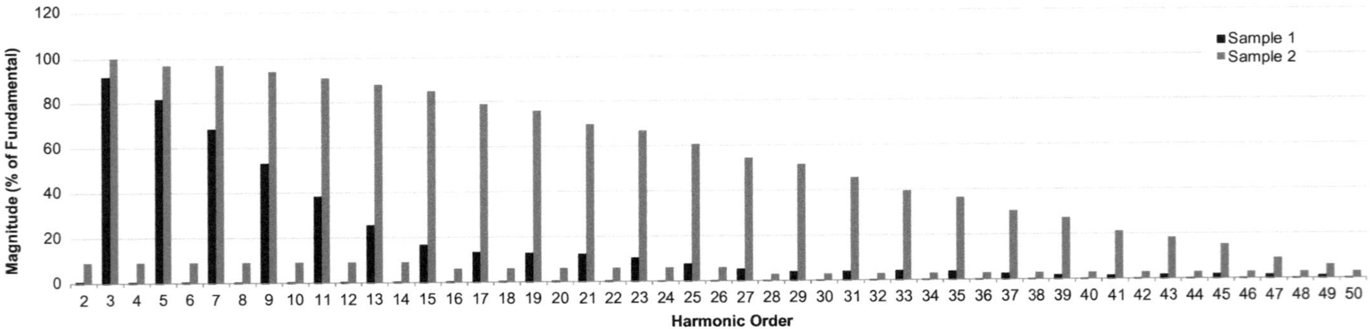

Fig. 7.57 Harmonic spectra for waveforms shown in Fig. 7.56.

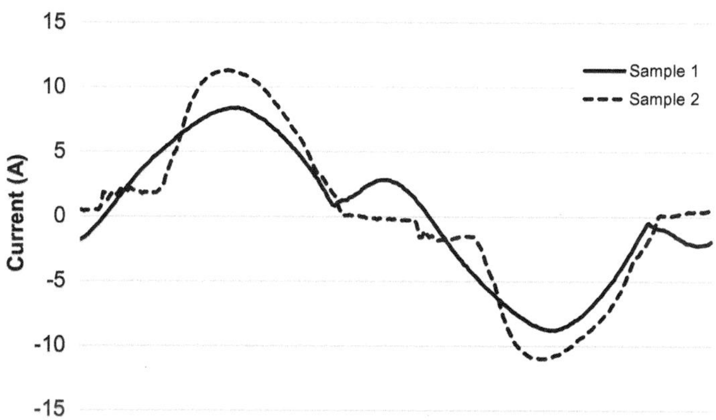

Fig. 7.58 Current waveforms for two sample microwave ovens.

5–7 MW (or larger). These inverters have much more sophisticated control systems to manage the power system dispatch and stability requirements of a large generator. Fig. 7.70 shows a photograph of two 2.5-MW solar farm inverters.

In terms of harmonic current emissions, these inverters have the following characteristics:

- Harmonic emissions are generally not in line with the 'traditional' emission profile of simple rectifier or inverter circuits. In many cases, the highest emission magnitudes occur at higher order harmonics.
- There is no strong relationship between emission magnitude and fundamental power. Highest emission (in amperes) may occur at lower output power.

Fig. 7.71 shows the current harmonic spectra for a 2.5-MW solar farm inverter for a number of operating (power output) points. There are a number of observations that can be made from these spectra:

- The dominant harmonic order is the 23rd.
- The maximum harmonic current magnitude (in amperes) for all measurements occurs at the lowest measured operating point, 750 kW.

Large inverters of this type, which will be connected at medium and high voltage, are dealt with very differently to the smaller inverters which are connected to low voltage networks. For these inverters there are no harmonic current emission limits for individual devices. Instead, it is likely that the installation will be specified an emission allocation by the network operator (Figs 7.70 and 7.71).

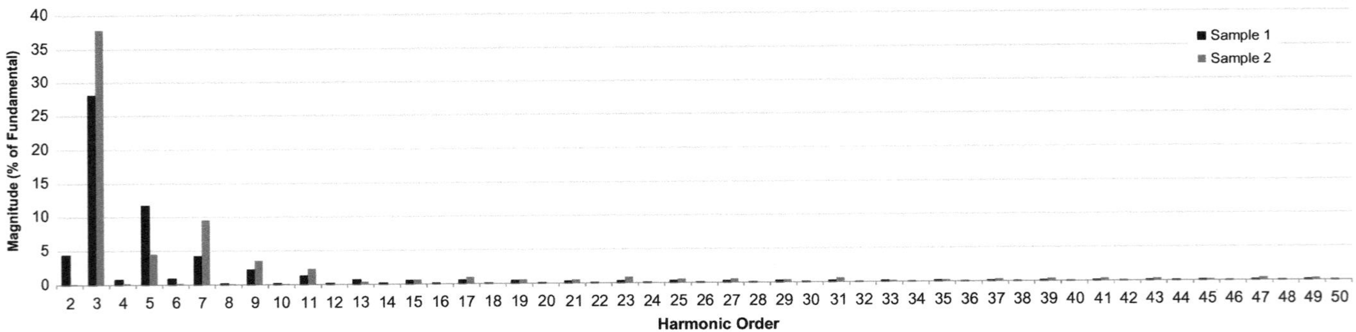

Fig. 7.59 Current harmonic spectra for two sample microwave ovens.

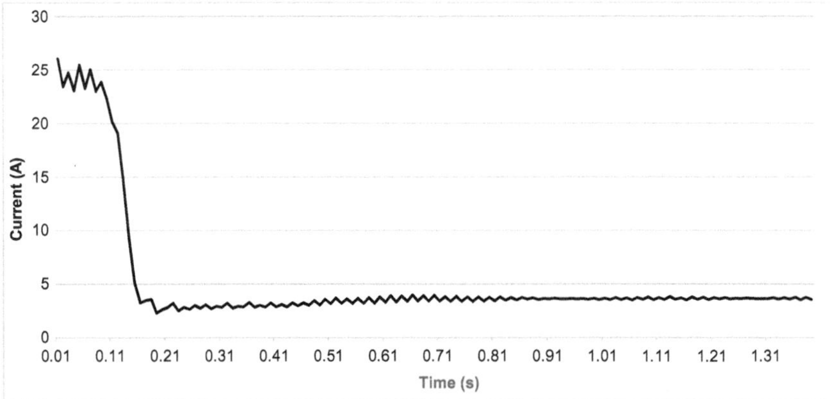

Fig. 7.60 Inrush current transient for starting of a direct-on-line air conditioner.

Fig. 7.61 Current waveform for inverter air conditioner.

7.7.6 Electric arc furnaces

Electric Arc Furnaces (EAFs) are used for the melting and refining of steel products and are a well-known source of flicker. The most prevalent use of EAFs is in the recycling of scrap steel, however, EAFs are also used in the production of high-grade alloy steel, aluminium, copper, lead and other metals.

The delivery of electrical energy to the scrap metal within an EAF is achieved via three graphite electrodes, which are lowered into the furnace. Each electrode is connected to one phase on the secondary side of the

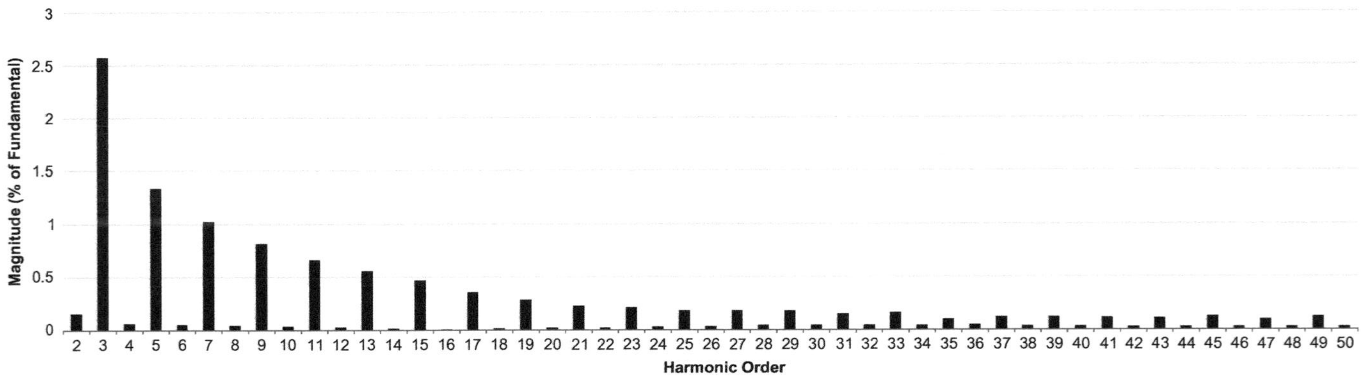

Fig. 7.62 Current harmonic spectrum for inverter air conditioner.

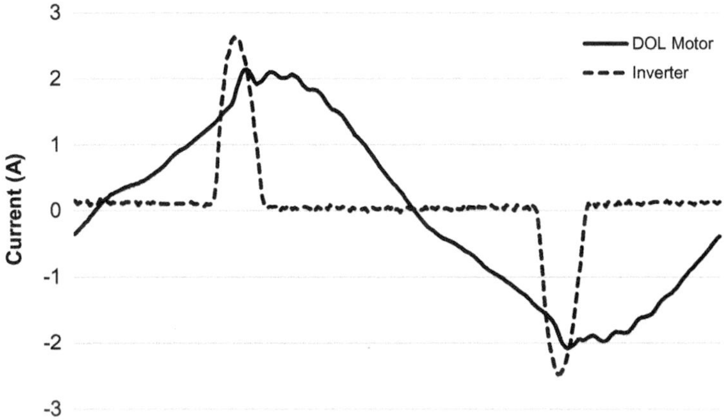

Fig. 7.63 Current waveform for two different refrigerator technologies.

furnace transformer via a flexible cable. Many large furnace transformers are rated at 100 MVA or greater. In order to begin the initial phase of the melting process, which is called the bore-in period, the electrodes are lowered one by one until they contact with the scrap metal. When the first electrode contacts with the scrap, the voltage between the scrap and ground becomes zero. When the next electrode makes contact, a high current passes from the first electrode to the second, through the scrap, causing an arc to form between the electrode and the scrap. Once an arc is established the electrodes are withdrawn and the arc is maintained by control mechanisms. Whilst the first electrode is being withdrawn, the third electrode is still being lowered, until it too strikes the scrap and an arc is set up, there is then a period where all three phases are conducting.

The temperature in the core region of an arc is between 10,000°C and 20,000°C and is easily enough to melt the metal. As stated, once an arc is established, the electrodes are withdrawn and the arc maintained by changing the position of the electrodes using hydraulics and PLCs.

The bore-in period is a period of extreme instability and is the source of many power quality problems. During the bore-in period, rapid movements of the electrodes are observed as the control systems strive to maintain a stable arc. The problems encountered in maintaining the arc are due to the fact that the surface of the material in the furnace is not uniform, meaning that in some cases all three phases will not be conducting at the same time and in other cases the arc may extinguish. Added to this is the fact that as the metal melts it will fall in on itself, creating randomly changing arc lengths and

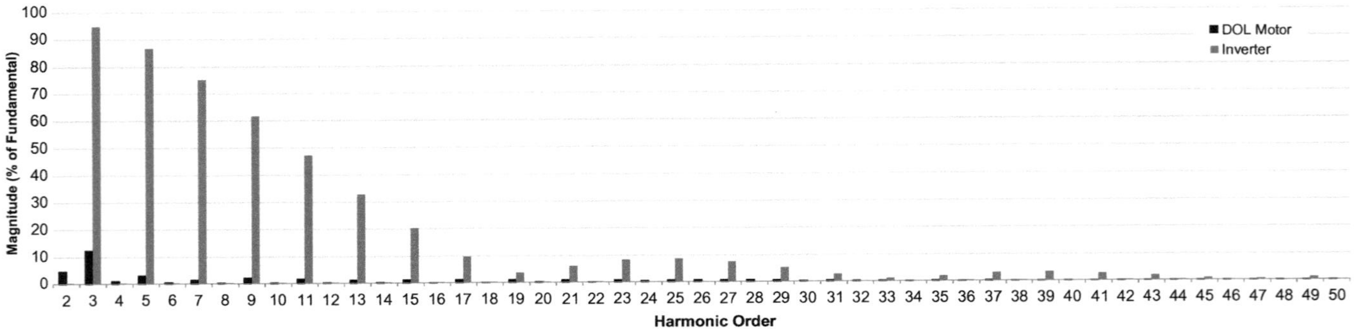

Fig. 7.64 Current harmonic spectrum for two different refrigerator technologies.

Fig. 7.65 Photograph of 20-kW solar PV inverters.

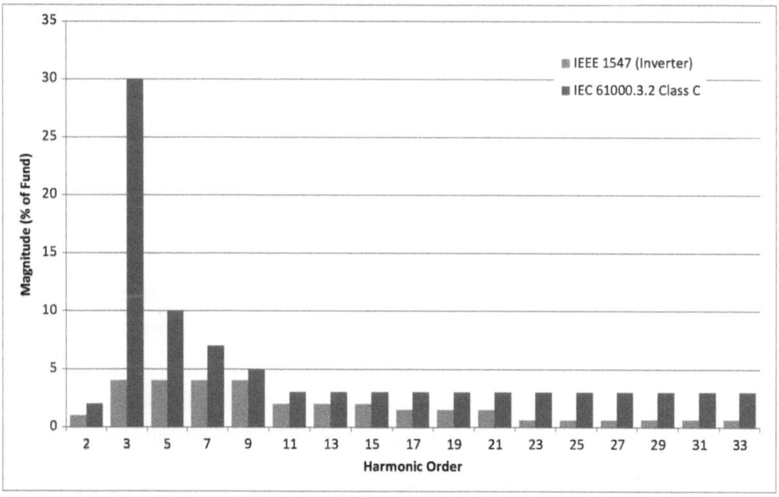

Fig. 7.66 Comparison of allowable harmonic emissions—renewable energy inverter compared to IEC 61000-3-2 Class C.

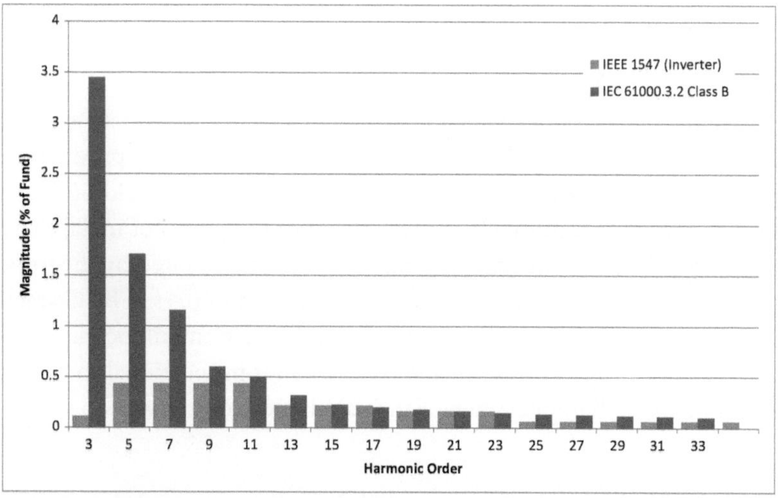

Fig. 7.67 Comparison of allowable harmonic emissions—renewable energy inverter compared to IEC 61000-3-2 Class B.

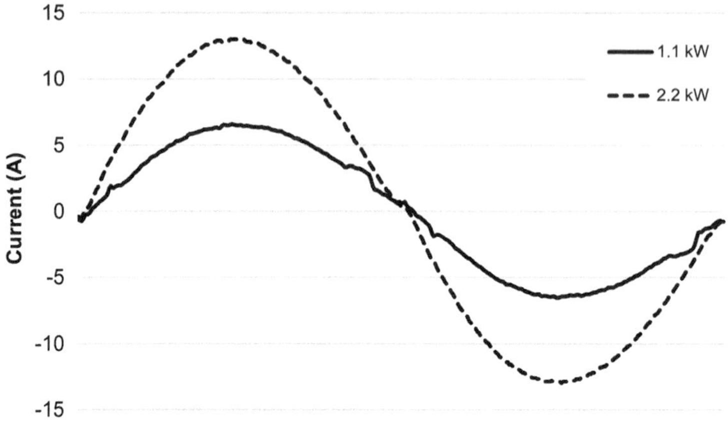

Fig. 7.68 Current waveform for 2.5-kW solar inverter.

possible arc extinction. These factors mean that during the bore-in period the arc is extremely erratic and unstable. The rapid movement of the electrodes means rapid changes in arc length and wide and chaotic swings in current are observed. These current fluctuations cause variations in reactive power, which cause a momentary voltage drop or flicker both at the supply bus and nearby buses in the interconnected system.

Fig. 7.72 shows the short-term flicker (P_{st}) trend measured in 10-min intervals across 24 h for an arc furnace load. Measurements are shown for the 22 and 132 kV buses at the site. The figure demonstrates the very high and very sporadic flicker emissions due to operation of the arc furnace. The figure also demonstrates the significant attenuation of flicker magnitudes that occurs as voltage level increases.

7.7.7 Mesh welders

Mesh welders are another well-known source of voltage fluctuations and flicker. Mesh welders use electric arc welding to manufacture products such as the steel reinforcing used in concrete slabs. In electric arc welding the supply is directly and intentionally shorted through the pieces of metal that are to be joined. This process draws a current that is only limited by the impedance of the supply. A mesh weld can be completed in 0.4 s. The characteristics of mesh welders are such that a voltage fluctuation will occur for each weld. The magnitude of this voltage fluctuation is proportional to the weld current as well as the impedance of the supply network. Fig. 7.73 shows an

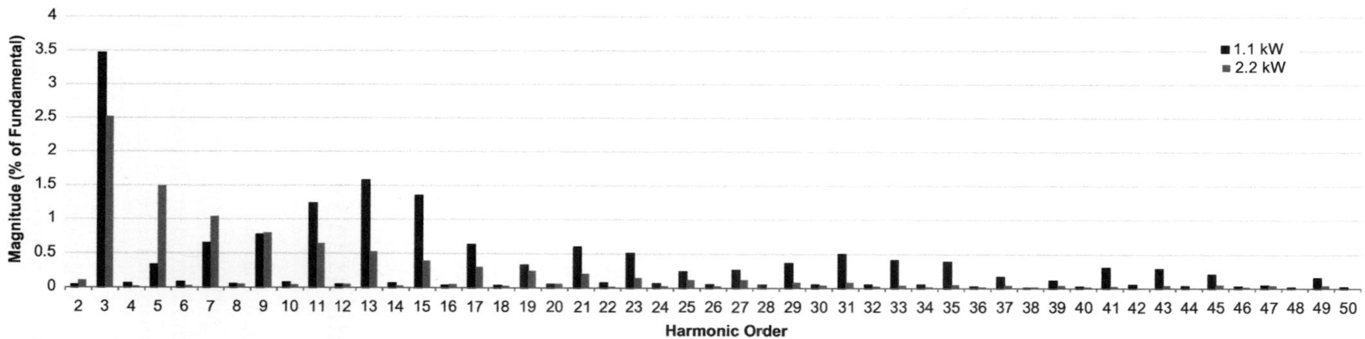

Fig. 7.69 Current harmonic spectra for 2.5-kW solar inverter.

Fig. 7.70 Photograph of solar farm inverters (2.5 MW)—one inverter at each end of image.

example of the RMS current and voltage for a mesh welder. In this figure, the short-term repetitive nature of the large weld current can be observed along with the impact on voltage. In this example, there is a 4% voltage change in voltage each time a weld is performed. Fig. 7.74 shows the impact of the mesh welder on the voltage waveform. The amplitude modulation of the waveform envelope associated with true flicker can clearly be observed. For this example, there are 214 voltage changes per minute. It can be calculated that this equates to a P_{st} value of approximately 8 (Figs 7.73 and 7.74).

7.8 Impact of variations in supply voltage on appliance performance

It is reasonably well known that variations in the supply voltage will cause variations in equipment performance. For example, waveform distortion may cause variations in the harmonic emissions of appliances. This section presents empirical (laboratory based) evaluation of the impact of variations in supply voltage magnitude as well as supply voltage harmonic distortion in order to provide the reader some understanding of the implications of variations in the supply voltage on appliance electrical behaviour.

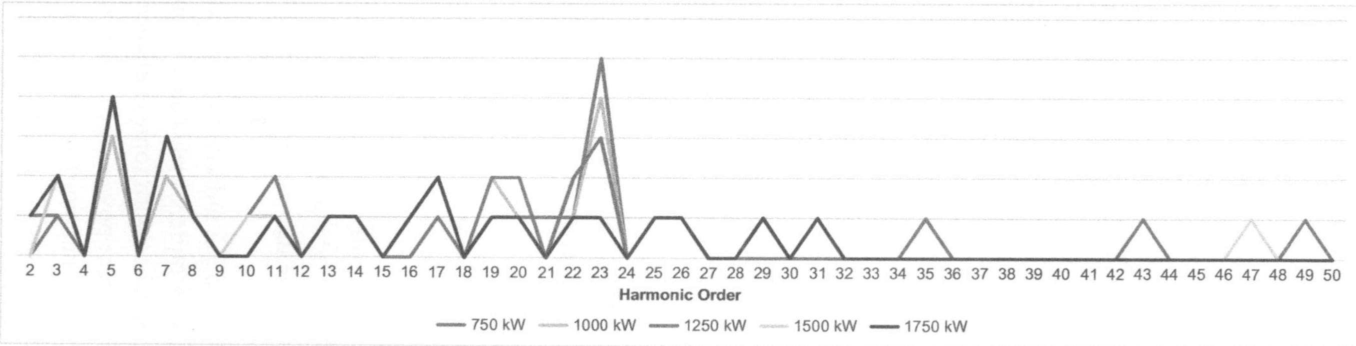

Fig. 7.71 Example current harmonic spectra for 2.5-MW solar farm inverter.

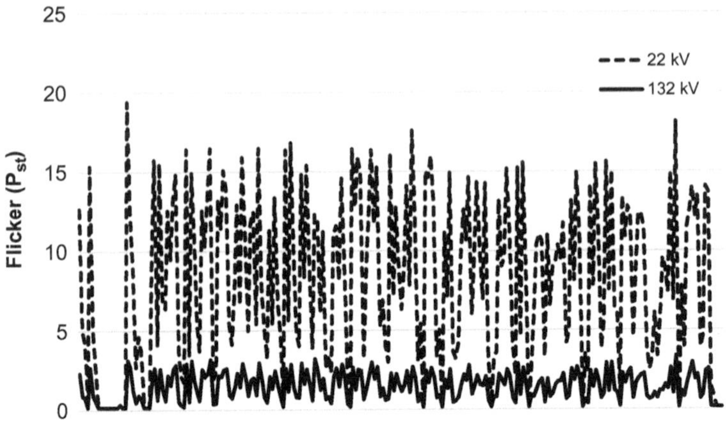

Fig. 7.72 One day flicker (P_{st}) trend for arc furnace load.

7.8.1 Impact of variation in supply voltage magnitude

The impact of variations in the magnitude of the supply voltage on the electrical performance for a range of common appliances is examined in this section. Understanding the implications of variation in the electrical performance of appliances due to changes in the supply voltage magnitude is essential for complete understanding of the manner in which appliances may interact with the electricity distribution system.

In order to illustrate the impact of supply voltage magnitude on key electrical parameters, data is presented for the case where appliances were supplied at 253-V RMS which is at the upper end of the allowable range (for a 230-V nominal system) and 207-V RMS which is at the lower end of the allowable range. Table 7.1 presents the outcomes (compared to the 230-V case) when appliances are supplied at 253-V RMS for the key parameters of RMS current (I RMS), displacement power factor (DPF), fundamental current (I Fund) and total harmonic current.

Table 7.2 presents the same data as Table 7.1 for a supply voltage magnitude of 207-V RMS.

The data in Tables 7.1 and 7.2 indicate that both the RMS current and fundamental current can vary quite widely based on supply voltage magnitude depending on the appliance with some appliances drawing more current as voltage reduced (constant power type operation) and other drawing less (constant impedance type operation). It can be seen that variation in supply voltage does not tend to have a significant impact on displacement power factor.

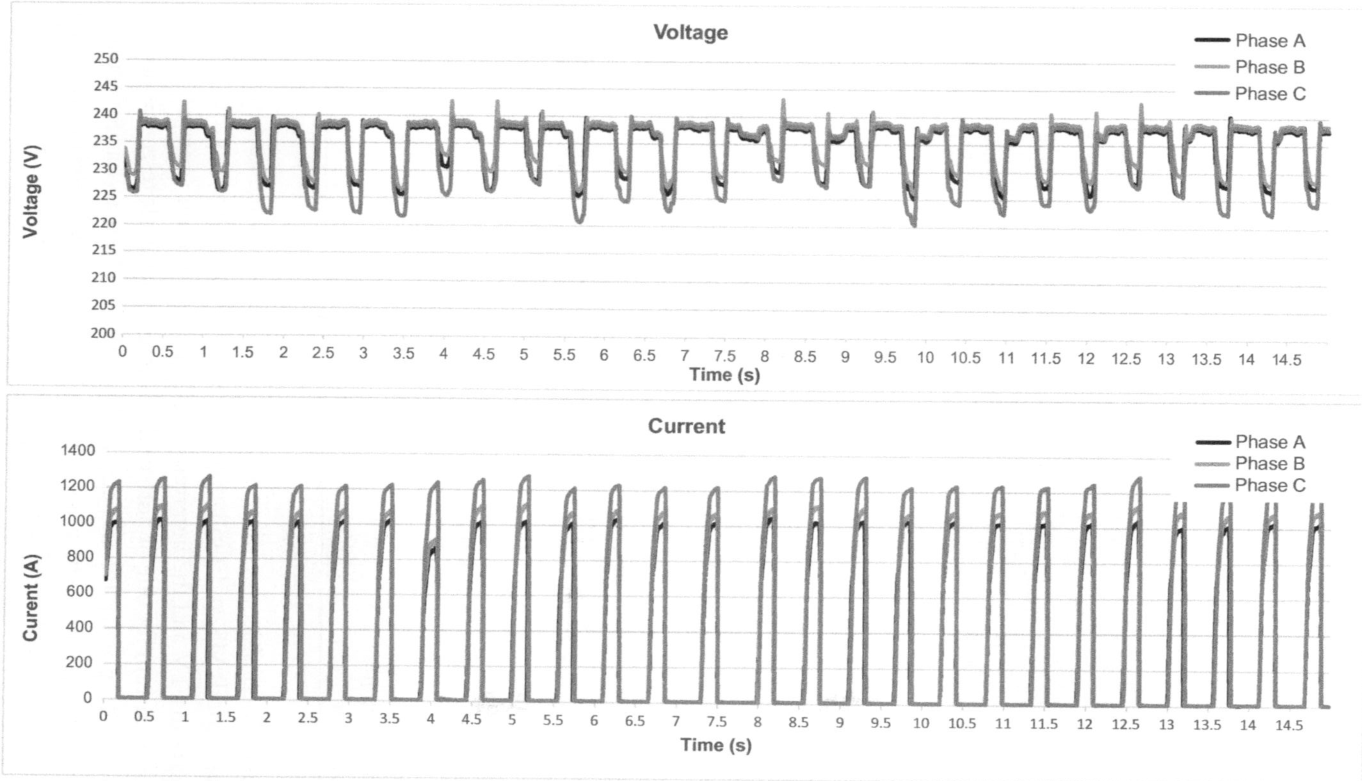

Fig. 7.73 Example of mesh welder operation (top: RMS voltage, bottom: RMS current).

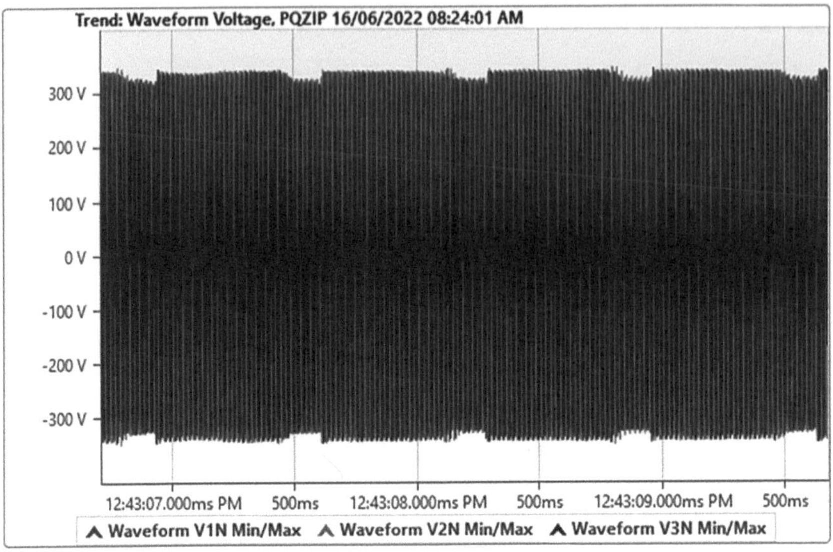

Fig. 7.74 Impact of mesh welder on voltage waveforms.

Table 7.1 Variation in key parameters when supplied at 253-V RMS (all values in percentage of value obtained when appliance supplied at sinusoidal 230-V RMS).

Appliance	I RMS	DPF	I Fund	Total harmonic current
CFL Sample 1	100	99	103	98
CFL Sample 2	100	98	102	97
CFL Sample 3	106	100	101	140
CRT television	107	100	111	106
LCD television	110	103	111	98
Plasma TV	102	101	104	88
DVD player	106	100	111	103
Inverter air conditioner	111	100	111	88
Personal computer	104	100	108	103
LCD monitor	105	100	110	103

There is considerably more variation in total harmonic current than is observed for RMS and fundamental current. For example, CFL Sample 3, the plasma TV and the inverter AC all show significant variation in total harmonic current for one or both of the voltage magnitude tests. It is notable that the CFL, plasma TV and air conditioner all appear to have electronic circuits as their input which aim to mitigate current waveform distortion. It may be postulated that these circuits are more sensitive to changes in input supply voltage than more simple devices such as the SMPS of the PC.

Table 7.2 Variation in key parameters when supplied at 207-V RMS (all values in percentage of value obtained when appliance supplied at sinusoidal 230-V RMS).

Appliance	I RMS	DPF	I Fund	Total harmonic current
CFL Sample 1	105	100	101	105
CFL Sample 2	102	101	100	105
CFL Sample 3	96	100	100	52
CRT television	92	100	88	93
LCD television	93	96	92	106
Plasma TV	105	100	103	121
DVD player	99	100	93	98
Inverter air conditioner	89	100	89	112
Personal computer	94	99	93	95
LCD monitor	95	100	91	96

Table 7.3 Variation in low order individual harmonic orders when supplied at 253- and 207-V RMS (all values in percentage of value obtained when appliance supplied at sinusoidal 230-V RMS).

	253-V RMS			207-V RMS		
Appliance	I3	I5	I7	I3	I5	I7
CFL Sample 1	99	96	92	102	104	107
CFL Sample 2	99	93	95	103	110	113
CFL Sample 3	162	122	160	12	63	29
CRT TV	110	109	107	88	89	91
LCD TV	97	109	104	101	99	91
Plasma TV	92	107	57	117	104	75
DVD player	110	109	107	93	93	94
Inverter AC	87	85	85	113	119	115
PC	107	106	105	92	93	93
LCD monitor	109	107	105	91	92	94

Table 7.3 presents the impact of varying the supply voltage magnitude on selected individual low order harmonic current values (3rd, 5th and 7th).

Table 7.3 shows that there is considerably more variation in individual harmonic current components than was observed for basic current parameters such as RMS current. Those appliances which showed the greatest variation in total harmonic current are also seen to have the most extreme variation in individual harmonic current components. The foremost of these is CFL Sample 3 whose low order current harmonic behaviour shows extreme sensitivity to the input voltage magnitude.

Table 7.4 Details of waveforms used to assess the impact of supply voltage distortion.

Harmonic test waveforms				
	Harmonic test waveform 1		Harmonic test waveform 2	
Harmonic order	Magnitude (% of fund)	Phase angle (°)	Magnitude (% of fund)	Phase angle (°)
3	1	0	2	0
5	3	180	4	180
7	1	50	1.5	50

7.8.2 Impact of variation in supply voltage distortion level

An assessment of the impact of variation in the distortion of the supply voltage waveform is based on appliance behaviour when supplied with two test waveforms with different levels of harmonic distortion. The details of the test waveforms are provided in Table 7.4. The fundamental voltage applied for both waveforms was 230-V RMS.

Tables 7.5 and 7.6 present the variation in key parameters for the two supply waveforms containing harmonic distortion.

Table 7.5 indicates that there is little variation in displacement power factor and fundamental current when the appliances are supplied with Harmonic Test Waveform 1. However, there is considerable variation seen for RMS current and total harmonic current. Further, the variations seen here are much larger than those observed in the previous section where input voltage magnitude was varied. This indicates that appliance behaviour

Table 7.5 Variation in key parameters when supplied by harmonic test waveform 1 (all values in percentage of value obtained when appliance supplied at sinusoidal 230-V RMS).

Appliance	I RMS	DPF	I Fund	Total harmonic current
CFL Sample 1	99	97	101	97
CFL Sample 2	101	99	98	104
CFL Sample 3	101	99	102	99
CRT TV	67	100	90	59
LCD TV	100	101	100	111
Plasma TV	117	100	119	102
DVD player	75	100	103	67
Inverter AC	105	100	105	147
PC	78	101	101	71
LCD monitor	70	100	99	60

Table 7.6 Variation in key parameters when supplied by harmonic test waveform 2 (all values in percentage of value obtained when appliance supplied at sinusoidal 230-V RMS).

Appliance	I RMS	DPF	I Fund	Total harmonic current
CFL Sample 1	104	98	99	104
CFL Sample 2	105	99	99	110
CFL Sample 3	105	98	102	131
CRT TV	79	100	96	73
LCD TV	103	101	103	118
Plasma TV	122	101	124	96
DVD player	87	101	102	83
Inverter AC	105	100	105	176
PC	87	101	99	84
LCD monitor	80	94	98	75

is more sensitive to changes in input voltage distortion as opposed to changes in input voltage magnitude.

Similar results are observed in Table 7.6 to those seen in Table 7.5. Variation in displacement power factor and fundamental current is modest; however, variation in RMS current and total harmonic current is significant.

Table 7.7 presents the variation in low order individual harmonic current orders when harmonic test waveforms 1 and 2 are used to supply the appliances under test. It can be seen that there is considerable variation in all harmonic orders for most appliances.

Table 7.7 Variation in low order individual harmonics when supplied by harmonic test waveform 2 (all values in percentage of value obtained when appliance supplied at sinusoidal 230-V RMS).

	Harmonic test waveform 1			Harmonic test waveform 2		
Appliance	I3	I5	I7	I3	I5	I7
CFL Sample 1	86	68	86	70	56	119
CFL Sample 2	86	91	185	78	101	237
CFL Sample 3	113	59	127	120	69	166
CRT TV	80	63	41	70	35	55
LCD TV	87	191	85	75	226	116
Plasma TV	97	228	75	81	281	74
DVD player	89	62	30	70	21	55
Inverter AC	69	326	116	64	404	157
PC	95	83	66	86	72	75
LCD monitor	88	66	37	70	34	56

7.9 Power quality standards for equipment performance

As a general principle, power quality levels in low voltage networks are managed by application of emission limits to individual items of equipment, that is, individual low voltage installations are not subject to a power quality allocation. Referring back to Chapter 1, the rationale here is that, provided that a comprehensive suite of power quality standards is applied, if equipment emissions are restricted to defined values, in most cases voltage disturbance compatibility limits can be maintained.

The IEC 61000-3 series of standards limit the emission of power quality disturbance levels from equipment connecting to low voltage networks. Standards exist for:

- Harmonic current distortion emissions
- Voltage fluctuations and flicker emissions

7.9.1 Harmonic current distortion

The IEC has two standards which specify limits for harmonic current emission for equipment connected to low voltage networks. These are as follows:

- IEC 61000-3-2 [2] which applies to equipment with an input current up to 16 A per phase.
- IEC 61000-3-12 [3] which applies to equipment with input current >16 A and ≤75 A per phase.

7.9.1.1 IEC 61000-3-2

IEC 61000-3-2 classifies equipment into 4 classes, A, B, C and D, and specifies limit for each class.

All equipment that cannot be classified as class B, C or D is classified as class A. Examples of class A equipment include:

- Balanced three-phase equipment
- Household appliances, excluding equipment identified as classes B, C or D
- Tools, excluding portable tools
- Dimmers for incandescent lamps
- Audio equipment
- Vacuum cleaners

The harmonic current emission limits for class A equipment are specified as absolute magnitudes of current (in amperes) for each harmonic order up to the 40th. The emission limits can be characterised as follows:

- Allowable emission decreases as harmonic order increases
- The limits for even harmonic order emissions are significantly lower than those allowed for odd harmonic orders

Equipment classified as class B is as follows:

- Portable tools
- Arc welding equipment which is not professional equipment

The emission limits for class B equipment are 1.5 times the values for class A equipment.

Class C equipment consists of lighting equipment. The limits for lighting equipment vary depending on the lamp technology and power rating (ratings above and below 25 W). In the case of lighting equipment with a rating between 5 and 25 W, the standard allows assessment based on either power-related limits or assessment based on what is effectively 'wave shape' criteria. It is this wave shape criteria that effectively results in the almost triangular wave shape as exemplified by the waveform of Sample 3 in Fig. 7.39.

Class D equipment is defined as equipment having a specified power (as measured according to the standard) of less than or equal to 600 W, of the following types:

- Personal computers and personal computer monitors
- Television receivers
- Refrigerators or freezers incorporating variable speed drives to control compressors

For the following categories of equipment, limits are not specified in the standard:

- Equipment with a rated power of 75 W or less
- Lighting equipment with a power rating less than 5 W
- Professional equipment with a total rated power greater than 1 kW
- A range of independent phase control with a rated power greater than 1 kW.

IEC 61000-3-2 also contains the methodology to be used to perform type testing of equipment for compliance with the standard.

7.9.1.2 IEC 61000-3-12

IEC 61000-3-12 specifies limits for equipment with input current 16 A and $\leq$75 A per phase to be connected to low voltage networks. The harmonic emission limits for equipment specified in IEC 61000-3-12 are presented as a percentage of the reference current (effectively the fundamental current) and vary based on the minimum short-circuit ratio of the location where the equipment is to be connected, termed R_{sce} in the standard. In simple terms, five possible minimum R_{sce} values are given:

- 33
- 66
- 120
- 250
- $\geq$350

The allowable harmonic emissions increase as R_{sce} increases. If a piece of equipment cannot meet the limits specified for an R_{sce} equal to 33 (i.e. the lowest emission limits) then the manufacturer is obliged to identify the R_{sce} value at which the equipment will comply with the standard and include that value in product documentation.

7.9.1.3 Requirements for connection of equipment at voltages above low voltage

Whilst IEC 61000-3-2 and IEC 61000-3-12 specify emission limits for individual pieces of equipment connecting to low voltage networks, equivalent standards do not exist for equipment to be connected to medium and high voltage networks. Instead, the harmonic emissions from equipment at these voltage levels are managed on an installation by installation basis with IEC TR 61000-3-6 (Electromagnetic compatibility (EMC)—Part 3-6: Limits—Assessment of emission limits for the connection of distorting installations to MV, HV and EHV power systems) [4] being an example of how this process can be implemented. In simple terms, each installation should be reviewed by the network operator and allocated emission limits for each harmonic order such that voltage planning limits can be maintained. The onus is then on the installation to ensure that their emission meets the allocation (and to take undertake mitigation if this is not the case). A detailed description of the management of harmonic in medium and high voltage networks is provided in Chapter 4.

7.9.2 Flicker and voltage fluctuations

For voltage fluctuations and flicker, there are two IEC standards which apply to equipment connecting to low voltage networks:

- IEC 61000-3-3 [5] which applies to equipment with rated current $\leq$16 A per phase; and
- IEC 61000-3-11 [6] which applies to equipment with rated current $\leq$75 A

In both of these standards, low voltage is defined as voltage magnitude between 220- and 250-V line to neutral voltage at 50 Hz.

IEC 61000-3-3 establishes test procedures (experimental, computer simulation or analytical) to determine the voltage fluctuation and flicker emissions of an individual piece of equipment and specifies limits for each. The experimental test procedure is predicated on the use of a reference impedance to evaluate voltage changes due to equipment operation. This reference impedance is intended to simulate the '95th percentile' impedance of a low voltage distribution network.

IEC 61000-3-11 establishes two criteria for equipment with higher current ratings:

1. If the equipment can meet the requirements of IEC 61000-3-3 it is deemed compliant.
2. For equipment that can't meet the requirements of IEC 61000-3-3, the manufacturer must specify either the maximum supply impedance to which the equipment can be connected in order to meet the requirements of IEC 61000-3-3 or declare that the equipment is only suitable for use in installations with a supply capacity ≥100 A per phase.

7.9.2.1 Requirements for connection of equipment above low voltage

Similar to the case for harmonics, whilst IEC 61000-3-3 and IEC 61000-3-11 specify emission limits for individual pieces of equipment connecting to low voltage networks, equivalent standards do not exist for equipment to be connected to medium and high voltage networks. Instead, voltage fluctuations and flicker emissions from equipment at these voltage levels are managed on an installation by installation basis with IEC TR 61000-3-7 (Electromagnetic compatibility (EMC)—Part 3-6: Limits—Assessment of emission limits for the connection of fluctuating installations to MV, HV and EHV power systems) [7] being an example of how this process can be implemented. Once again, each installation should be reviewed by the network operator and allocated emission limits for each harmonic order such that voltage planning limits can be maintained. The onus is then on the installation to ensure that their emission meets the allocation (and to take undertake mitigation if this is not the case). A detailed description of the management of harmonics in medium and high voltage networks is provided in Chapter 5.

References

[1] S. Elphick, P. Ciufo, S. Perera, The electrical performance of modern compact fluorescent lamps, Aust. J. Electr. Electron. Eng. 7 (1) (2010) 43–51.

[2] IEC, Electromagnetic Compatibility (EMC)—Part 3-2: Limits—Limits for Harmonic Current Emissions (Equipment Input Current ≤16 A per Phase), 2020.

[3] IEC, Electromagnetic compatibility (EMC)—Part 3-12: Limits—Limits for Harmonic Currents Produced by Equipment Connected to Public Low-Voltage Systems With Input Current >16 A and ≤ 75 A Per Phase, IEC 61000-3-12:2011, 2011.
[4] Standards Australia, Electromagnetic Compatibility (EMC)-Limits—Assessment of Emission Limits for the Connection of Distorting Installations to MV, HV and EHV Power Systems TR IEC 61000-3-6:2012, 2012.
[5] IEC, Electromagnetic Compatibility (EMC)—Part 3-3: Limits—Limitation of Voltage Changes, Voltage Fluctuations and Flicker in Public Low-Voltage Supply Systems, for Equipment With Rated Current ≤ 16 A Per Phase and Not Subject to Conditional Connection, IEC 61000-3-3:2013, 2013.
[6] IEC, Electromagnetic Compatibility (EMC)—Part 3-11: Limits—Limitation of Voltage Changes, Voltage Fluctuations and Flicker in Public Low-Voltage Supply Systems—Equipment With Rated Current ≤ 75 A and Subject to Conditional Connection, IEC 61000-3-11:2017, 2017.
[7] IEC, Electromagnetic Compatibility (EMC)—Part 3-7: Limits—Assessment of Emission Limits for the Connection of Fluctuating Installations to MV, HV and EHV Power Systems IEC TR 61000-3-7, 2008.

CHAPTER 8

Power quality monitoring, data analysis and reporting

8.1 Introduction

This chapter describes the processes for monitoring of power quality as well as data analysis and reporting techniques which can be applied to analyse and visualise power quality data. In the first instance, the concept of power quality monitoring is described, including the motivation for its undertaking. The role of standards in power quality monitoring is then discussed. A review of power quality disturbances is presented including the methodology that should be applied for their monitoring, analysis and reporting. Power quality instruments and transducers are then reviewed. A detailed description of the two motivations for power quality monitoring is outlined before the chapter concludes with a description of methods for power analysis and reporting of power quality data.

8.1.1 What is power quality monitoring?

Power quality monitoring is the assessment of power quality levels or disturbances present in an electricity supply network or within an installation. In Ref. [1] a power quality disturbance is defined as "any power problem manifested in voltage, current or frequency deviations that results in failure or misoperation of customer equipment." More simply, a power quality disturbance can be considered as an excursion from the power system sinusoidal nominal voltage and frequency.

It is important to note the distinction between power quality and reliability. Reliability, which is in fact a subset of power quality, refers strictly to the availability of electricity supply whilst the wider umbrella of power quality deals with the full spectrum of the quality of the electrical supply and the abnormalities which may occur. Power quality monitoring is undertaken through the use of specialised instrumentation designed to measure the various power quality disturbances.

Applied Power Quality
https://doi.org/10.1016/B978-0-323-85467-2.00005-6

8.1.2 Reasons for PQ monitoring

There are two distinct motivations for power quality monitoring. For the purposes of this chapter, these are termed reactive and proactive monitoring. The reasons for, and methodology of, conducting the monitoring required to satisfy each monitoring motivation can differ markedly. A short definition for each monitoring motivation is as follows:

- **Reactive monitoring** is conducted in response to a known abnormal operation of the electricity system often arising from either a customer complaint (in the case of the network operator) or maloperation or failure of equipment.
- **Proactive monitoring** refers to routine monitoring usually performed by network operators. Proactive monitoring can be used for a variety of reasons, including regulatory requirements and for planning purposes.

Table 8.1 presents a summary of the reasons or motivation for undertaking power quality monitoring.

8.1.3 Power quality monitoring timescale and amplitude aspects

Power quality disturbances have a wide range of frequency and magnitude characteristics. Table 8.2 presents the effect on voltage waveforms for some of the most common power quality disturbance types.

Where frequency effects are concerned, some disturbance types require long monitoring periods in order to produce discernible trends whilst others require very high sample rates in order to capture important information. Both requirements can impose a very high dataflow and memory burden on power quality monitoring instrumentation (monitors) unless sophisticated designs are used. For example, steady-state voltage varies slowly, potentially over a timescale of many minutes; at the other extreme, lightning transients (impulsive type) can have rise times of a microsecond. Long-term monitoring

Table 8.1 Possible reasons/motivations for undertaking power quality monitoring.

Investigation type	Reason/motivation
Reactive	• Investigate equipment problems in an installation or on the network • In response to customer complaints • Before connection of large loads
Proactive	• To obtain general PQ levels in a part of the network • To satisfy regulator requirements • To assist in planning and asset management decisions

Table 8.2 Common PQ disturbance effect on voltage waveforms.

Classification	**Waveform**
Steady-state voltage	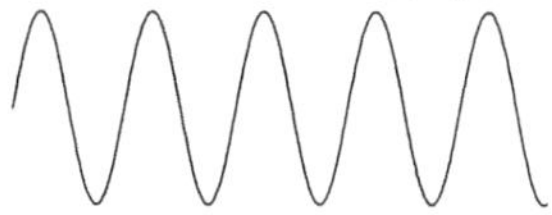
Short-term interruption/sag or dip	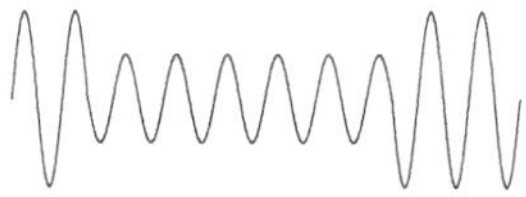
Voltage fluctuations	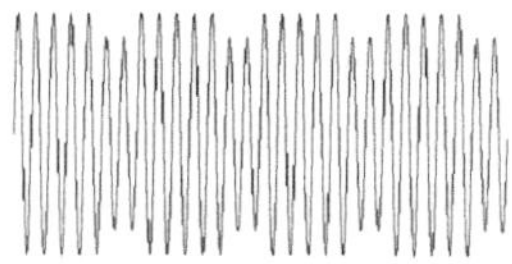
Harmonics	
Transients **(a)** Impulsive **(b)** Oscillatory	

is required to establish steady-state voltage trends whilst a high sampling rate is required to capture the transient phenomena. The range of timescales over which power quality disturbances tend to vary is presented in Table 8.3.

Similar considerations need to be made with respect to amplitude effects to those observed for frequency effects. Voltage fluctuations require a fine resolution whilst transients demand a large input voltage range (including crest factors to avoid clipping).

Table 8.3 Frequency range of PQ disturbances.

Disturbance type	**Timescale**	**Increasing frequency range**
Steady-state voltage	Minutes	
Unbalance	Minutes	
Harmonics	Seconds–minutes	
Short-term interruption	Seconds–minutes	
Sag	Cycles–minutes	
Swell	Cycles–minutes	
Voltage fluctuations	Cycles–minutes	
Transients	Subcycle	

8.1.4 Continuous and discrete disturbances—Monitoring and reporting requirements

Power quality disturbances can be classified into two main categories: continuous or discrete. This classification has an important bearing on how the disturbances are measured. Continuous disturbances can also be referred to as variations or quasisteady-state parameters. Discrete disturbances are sometimes referred to as events. Table 8.4 presents the classification (continuous or discrete) for the most common power quality disturbances.

Continuous disturbances such as steady-state voltage level are described by constant values under steady-state conditions (e.g. under laboratory conditions). In this situation they can be read using simple handheld metres with no data-logging capability. Under field conditions there will be time variation because of disturbing effects. When the variation is fast, handheld metres may be ineffective and it is preferable to continuously log the quantity. The quantity may then be represented by a time series trend graph, an example of which is shown in Fig. 8.1 which shows a plot of the value measured at regular intervals over time. Typically a 10-min measurement interval is used for continuous power quality disturbances.

If the quantity is to be characterised by a single number, a statistical measure such as average or maximum, also shown in Fig. 8.1, might be used. In

Table 8.4 Continuous and discrete disturbances.

Continuous disturbances (present in every cycle)	Discrete disturbances (present in a small fraction of cycles)
Voltage	Sag/interruption
Unbalance	Swell
Harmonics	Transient
Fluctuation	

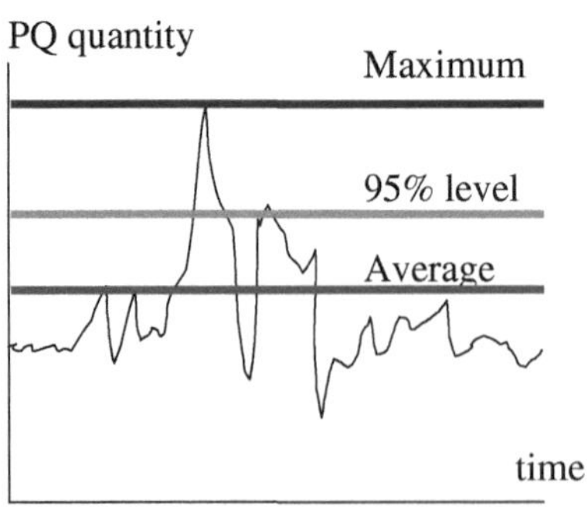

Fig. 8.1 Trend graph example.

practice, the average can be too optimistic since the waveform is larger than this for 50% of the time, often by a factor of two or more. However, the maximum has been found to be too pessimistic as it occurs only very briefly. An intermediate approach is to use the 95% value (sometimes called the 95% cumulative probability level or 95th percentile), the value which is not exceeded for 95% of the time. Note that the average is also the 50% value whilst the maximum is the 100% value.

Although the use of the 95% value is generally satisfactory, it will be exceeded for 8.4 h in a week. Use of the 95% value assumes that the behaviour in the remaining 5% of the week is not too extreme. If a situation arises where this is not so, other statistics may have to be considered.

Discrete disturbances such as transients or voltage sags usually occur as isolated disturbances over a few cycles with potentially long intervals before they are repeated. The concept of 'steady state' is not applicable to these disturbances since, if the term was to mean anything, it would be that they are absent. These disturbances cannot be examined by handheld metres without logging capability.

A monitoring instrument identifies the occurrence of a discrete disturbance by the presence of a cycle which is different to the cycle before it. However, if measurements were made with a small enough resolution, no two cycles would be the same. Hence the question of how different a cycle has to be before a discrete disturbance can be said to have occurred arises. The level of difference which is considered to be significant is called the threshold and is often adjustable in power quality monitors. Once the event threshold has been exceeded, the disturbance type needs to be recognised, its parameters determined and stored together with the time and date stamp, a process called event capture. For example, consider the voltage sag shown in Fig. 8.2, for the purposes of this example, assume that the sag

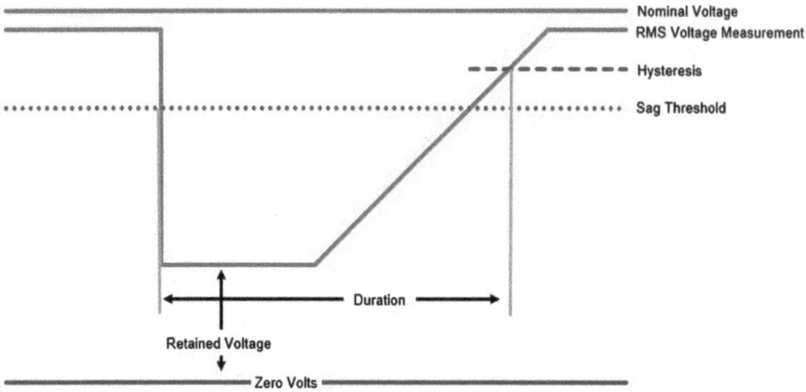

Fig. 8.2 Characterising a discrete disturbance—voltage sag example.

occurs at 15:31.06 on 1st January 2022. This sag has a voltage reduction of 35% (i.e. to 65% retained voltage) for 2 cycles (0.04 s). Depending on the instrument design, it may record 'Sag 2022:01:01:15:31.06 65 0.04'. A list of such events and their parameters gives an event log.

8.2 Standards for power quality monitoring

8.2.1 Role of standards

As with the measurement of any quantity, standards are required to ensure the repeatability and comparability of power quality measurements. There are three types of standards which need to be considered when conducting power quality monitoring:

(i) Standards which specify measurement techniques, algorithms and accuracies.

(ii) Standards which define how power quality instrumentation should be designed and how it should operate.

(iii) Standards which define acceptable disturbance levels.

Concerning (i), the internationally recognised standard which outlines the manner in which power quality instrumentation should operate is IEC 6100-4-30 [2]. The scope of IEC 61000-4-30 is to define the methods for measurement and interpretation of results for power quality parameters. Quoting from the standard, "Measurement methods are described for each relevant parameter in terms that will make it possible to obtain reliable, repeatable and comparable results regardless of the compliant instrument being used and regardless of environmental conditions" [2]. It is important to note that IEC 61000-4-30 is a performance specification, not a design specification, meaning that it does not specify precise instrument design details such as required sampling frequency; instead, it describes instrument performance requirements such as accuracy, measurement techniques and aggregation periods to ensure repeatability within specified accuracy constraints. The standard also provides informative guidelines on minimum assessment periods and recommended reporting intervals. IEC61000-4-30 is examined in detail in the subsequent sections.

Concerning (ii), there are a range of standards which defined the specifications and design requirements of instrumentation used to measure power quality. These include IEC 61000-4-15 [3] which is the design specification for the flickermeter which is used to measure flicker, and IEC 61000-4-7 [4] which defines the requirements for instrumentation used to measure harmonic distortion should operate. Where necessary, these standards are referenced by IEC 61000-4-30 to provide guidance on the measurement of certain disturbances.

The third (iii) type of standards relevant to power quality monitoring is standards which define acceptable power quality levels or planning limits. These standards are important as these allow assessment of the overall power quality 'health' of the network and allow planners to design to commonly accepted criteria. These standards outline acceptable limits or planning levels for various power quality disturbances. One of the major problems in many places is that in many cases there are no consistent power quality standards applied across all jurisdictions. For example, in Australia, whilst the National Electricity Rules apply in the interconnected national electricity market (NEM), each Australian state also has a regulator which defines power quality limits independently for their own jurisdiction. Added to this, each network operator may have their own codes or system performance standards defining acceptable power quality levels for their networks.

8.2.2 IEC 61000-4-30

Whilst standards concerning limits for various power quality disturbances have been in use for some time and standards such as EN50160 [5] and IEEE 1159 [6] provide detail on performing and evaluating power quality surveys, until the release of IEC 61000-4-30 in 2003 there was no standard which specifically and extensively described methods of conducting power quality surveys and measuring power quality disturbances in order to ensure comparable and repeatable measurements regardless of the instrument type or manufacturer. Prior to the release of IEC 61000-4-30, most measurement functions, including accuracy, sampling frequency and data aggregation methods, were at the discretion of the instrument manufacturer and were often driven by specific customer requirements rather than with a view to following a standardised measurement protocol. Consequently, before the widespread acceptance of IEC 61000-4-30 it was difficult to compare the results of a survey made with one type of instrument with those of another type of instrument.

Power quality disturbance measurement techniques covered by the normative sections of IEC61000-4-30 are as follows:

- voltage and current variation
- voltage unbalance
- frequency
- voltage dips and swells

In addition, the standard references IEC 61000-4-7 for harmonics and interharmonics measurement protocols and accuracies and IEC 61000-4-15 for flicker measurement protocols and accuracies. The only major power quality disturbance not covered by the standard is transients and to date there is currently no standardised method for measuring and evaluating transients.

The focus of IEC 61000-4-30 is to describe the way in which instruments should operate to produce standardised and repeatable measurement results. To achieve these outcomes, the standard outlines three classes of instrument: Class A, B and S. Class A instruments are very precise and comply with strict accuracy limits and methods of data sampling and aggregation. Class B instruments are of less interest here as their specifications are stated by the manufacturer (there are minimum performance standards). Class S which falls between Class A and Class B has requirements similar to Class A but generally requires less accuracy and less instrument processing power. This classification of instruments leads to one of the greatest benefits of the standard, namely, the standard ensures that any two instruments complying with Class A and measuring the same signal should record the same result within the strict tolerances of the standard.

8.2.2.1 IEC61000-4-30 instrumentation classes

The 2015 edition of IEC61000-4-30 defines three classes of measurement (instrumentation classes) as detailed as follows:

Class A: These instruments are used where very high accuracy and precise measurements are required. They are useful for situations with legal implications, contractual applications, verifying compliance with standards and resolving disputes.

Class S: These instruments are useful for statistical survey where very high accuracy is not necessary. They have similar requirements to Class A; however, accuracies and processing requirements are generally lower.

Table 8.5 presents a summary of the accuracy requirements for Class A and Class S instruments.

Class B: The accuracies and measurement algorithms for these instruments must be specified by manufacturer (there are extreme limits). This type of instrumentation is useful for statistical surveys and troubleshooting where very high accuracy is not required. Class B instruments may ultimately be removed from IEC61000-4-30; however, they remain in the standard at the moment to avoid rendering legacy instruments obsolete.

Table 8.5 Summary of accuracy requirements for Class A and Class S instruments.

Disturbance	Class A	Class S
Voltage variation	±0.1% (10%–150% of declared input voltage)	±0.5% (10%–150% of declared input voltage)
Residual voltage (sags) and swell voltage	±0.2% (10%–150% of declared input voltage)	±1% (10%–150% of declared input voltage)
Unbalance	±0.15% for both negative sequence ($u2$) and zero sequence ($u0$) components	
Voltage harmonics	$U_m \geq 1\%\ U_{nom}$:±5% U_m $U_m < 1\%\ U_{nom}$:±0.05% U_{nom}	$U_m \geq 3\%\ U_{nom}$:±5% U_m $U_m < 3\%\ U_{nom}$:±0.15% U_{nom}
	U_m is measured value, U_{nom} is nominal voltage range of instrument	

8.2.2.2 IEC61000-4-30 aggregation intervals

IEC 61000-4-30 specifies a range of aggregation intervals that should be used for the evaluation of power quality levels. These aggregation intervals apply only to Class A and Class S instrumentation. For Class B instrumentation, aggregation intervals are at the discretion of the instrument manufacturer. The IEC61000-4-30 aggregation intervals are presented in Table 8.6. These aggregation intervals are in turn derived from a number of shorter intervals. For 50-Hz systems the basic interval is 10 cycles. The methods by which the aggregation intervals shown in Table 2.1 are calculated from 10-cycle values is one of the major differences between Class A and Class S instrumentation. The requirements in terms of precision timing and retention of 10-cycle 'building blocks' for Class A are much more onerous than those for Class S. From these basic aggregation periods, longer evaluation periods may be calculated by averaging (arithmetic or RMS) and statistical measures may be used to evaluate power quality levels.

Table 8.6 IEC 61000-4-30 aggregation intervals.

Abbreviation	Name	Duration
vs	Very short	3 s (strictly 150/180 cycles)
s	Short	10 min
l	Long	2 h
d	Daily	Day
w	Weekly	Week
y	Yearly	Year

8.2.2.3 *Flagging concept*

The flagging concept specified in IEC61000-4-30 aims to prevent one disturbance from being reported more than once. For example, flagging attempts to avoid a sag (an event) being counted as a steady-state voltage variation (a continuous disturbance) as well as a sag. The flagging concept is only applicable to Class A instrumentation and states that during the survey period, if a discrete disturbance (event) is detected, the continuous data recorded during the interval in which the event was detected should be flagged. This flag allows quick identification of data that may not be a reliable indicator of steady-state performance.

Flagging as a concept is useful in preventing a single disturbance influencing other disturbances and to separate out discrete events from continuous data; however, application of the flagging concept may not be straightforward. This is especially true in the case of flicker. As an example, Fig. 8.3 shows a 1-week flicker time series graph (10-min intervals) obtained from monitoring of a zone substation auxiliary transformer. It can be seen that there are many flicker spikes rising to quite high P_{st} levels. When the event logs were interrogated it was found that each high flicker reading corresponded to a sag on the auxiliary transformer to which the power quality monitor was connected. Further investigation revealed that these sags were being caused by the starting of a ripple injection load control motor–generator (a load control technology) set that was supplied by the auxiliary transformer. This means that the flagging concept should be applied to each of the flicker readings corresponding to a time interval in which a sag occurred would be removed. This results in the following problems with assessment of the data:

- The first problem is that there are so many sags that the removal of the data corresponding to sags severely reduces the amount of data upon which analysis can be performed, thus affecting the statistical accuracy

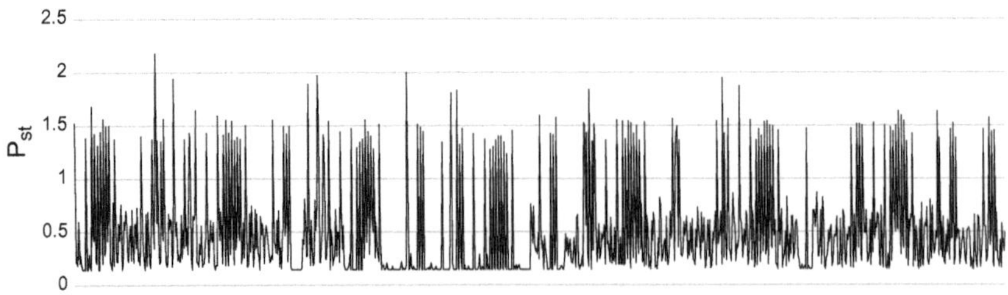

Fig. 8.3 One-week flicker time series graph for zone substation auxiliary transformer.

of the sample; however, at the same time, the data cannot be included as the high values are obviously caused by the sags.

- The second problem corresponds to the calculation of the long-term flicker index, P_{lt} which relies on 12 consecutive P_{st} values. Obviously with the loss of so much data it is virtually impossible to calculate P_{lt}. At the present time, IEC61000-4-30 is silent as to how these issues should be addressed.

This example demonstrates two important practical considerations which must be taken into account when power quality monitoring is performed. Firstly, it shows that placement of instrumentation can have a direct bearing on the quality of the data obtained, and secondly that application of the flagging concept is not straightforward.

8.3 PQ disturbances and their characterisation

This section describes the characteristics of the main power quality disturbances as well as the measurement and reporting methodology for each.

8.3.1 Steady-state voltage magnitude/variation

Definition: Steady-state voltage magnitude/variation is simply the RMS phase-to-phase or phase-to-neutral magnitude of the supply voltage.

Limits: As discussed in Chapter 2, there are multiple limits for steady-state voltage magnitude in effect internationally, including two main low voltage nominal voltages, 230 V and 110 V. There is also a wide variety of nominal voltages for medium and high voltage networks.

Measurement: According to IEC61000-4-30, for Class A instrumentation, "the measurement shall be the RMS value of the voltage magnitude over a 10-cycle time window for 50-Hz power system or 12-cycle window for 60Hz power system. Each 10/12-cycle interval shall be contiguous and non-overlapping" [2]. These 10/12-cycle values may then be RMS averaged to a larger time period with 10 min being widely applied and internationally accepted.

Assessment and reporting: Voltage magnitude is not technically a disturbance and is the only common power quality quantity for which the optimum value should not ideally be zero (for example the value result for disturbances such as voltage unbalance or total harmonic distortion is zero). Thus a method of voltage reporting is required that creates a condition whereby zero is the optimum result and increasing values indicate increasing levels of disturbance.

IEC TS 62749 [7] provides recommended values for power quality indices. In the case of voltage, indices to describe performance at the upper and lower ends of the voltage range are defined. For the upper end of the range, indices such as the maximum (100%) and 99th percentile are suggested. For the lower end of the range, 5th percentile, 1st percentile or minimum values are suggested. It is recommended that assessment of performance be based on 10-min RMS values collected over a period of 1 week.

In Australia, AS/NZS 61000.3.100 prescribes assessment of voltage levels to be undertaken by comparing the 99th percentile and 1st percentile voltage levels calculated using 10-min values collected over a 1-week period to the applicable limit.

8.3.2 Unbalance

Definition: In Ref. [1], unbalance is defined as "a condition in which the 3 phase voltages differ in amplitude or are displaced from their normal 120 degree phase relationship or both."

There are two definitions of unbalance factor, the international IEC definition and a US IEEE/NEMA definition. The IEC definition is becoming widely adopted and is given here. If a three-phase set of voltages has positive and negative sequence components V_P and V_N,

$$\text{VUF (Voltage Unbalance Factor)} = V_N/V_P \tag{8.1}$$

This definition does not take into account zero sequence because it has no adverse effect on loads and is often absent anyway. Perhaps a better name for VUF is 'negative sequence unbalance factor'. Voltage unbalance factor can also be calculated from simple line–line readings using the following equations:

$$\beta = \frac{V_{ab}^4 + V_{bc}^4 + V_{ca}^4}{\left(V_{ab}^2 + V_{bc}^2 + V_{ca}^2\right)^2} \tag{8.2}$$

$$\text{VUF} = \sqrt{\frac{1 - \sqrt{3 - 6\beta}}{1 + \sqrt{3 - 6\beta}}} \tag{8.3}$$

Limits: IEC 61000.3.13 [8] specifies a compatibility level of 2% for voltage unbalance. This compatibility level is often used as a limit in low voltage networks. IEC 61000.3.13 also specifies the indicative planning levels presented in Table 8.7 for unbalance in MV, HV and EHV systems. These planning level values are often used as limit in these networks.

Table 8.7 Indicative planning levels for negative sequence unbalance in MV, HV and EHV networks from Ref. [8].

Voltage level	Planning level (%)
MV	1.8
HV	1.4
EHV	0.8

Measurement: Where instruments are able to measure and calculate unbalance, "the fundamental component of the RMS voltage input signal is measured over a 10-cycle time interval for 50Hz power system" [2]. The effects of harmonics should be minimised by the use of a filter or by using DFT algorithm. These 10-cycle values may then be RMS averaged to a larger time period with 10 min being internationally accepted.

It should be noted that, wherever possible, measured unbalance should be determined from 10-cycle RMS voltage values averaged to a 10-min time interval. Different instruments may use different methods to calculate unbalance and it is important to ensure that unbalance is being calculated correctly. Failure to sample correctly for unbalance calculation may lead to the loss of high-frequency unbalance data, and unbalance calculated in this manner has been observed to be up to 30% lower than correctly calculated unbalance [9].

As described before, where directly measured unbalance is unavailable, unbalance may be calculated using the three corresponding voltages and time periods by way of the IEC formula shown in Eq. (8.3). Again there is the problem of sampling periods and this should be taken into account when unbalance values are examined. In addition, the IEC formula requires use of line-to-line voltages. Where line-to-line voltages are not available, line-to-neutral voltages may be used and substituted in the calculation. The use of line-to-neutral voltage in this calculation can result in additional zero sequence components, but these are usually small when 10-min values are used [9].

Reporting: The commonly accepted primary reporting index for unbalance is the weekly 95th percentile value of unbalance measurements collected using 10-min recording intervals.

8.3.3 Harmonics

Definition: "Harmonic distortion is the change in the waveform of the supply voltage from the ideal sinusoidal waveform. It is caused by the interaction of distorting loads with the supply network" [10].

Limits: In most jurisdictions, electricity rules or regulations require that network operators maintain voltage harmonic levels below defined compatibility levels. To achieve this the network operator must adopt harmonic planning levels below the compatibility levels and design a harmonic emission allocation strategy to ensure this if these planning levels are not exceeded. The magnitude of the planning levels is generally at the discretion of the network service provider. Indicative values for planning levels for harmonics are provided in IEC 61000-3-6 and IEEE 519 [11].

Measurement: The voltage or current waveform has to be sampled many times a second, depending on how many harmonics are required to be determined. The FFT (Fast Fourier Transform) is then applied to find the fundamental component (V_1 or I_1) and harmonics (V_2/I_2, V_3/I_3, and so on) usually to the 40th or 50th order. These values are averaged over 10-min intervals. The total harmonic distortion (THD) is determined at the same time.

Sometimes it is useful to determine harmonic phase angle as well. This enables the harmonic power to be determined. Its direction can be useful in identifying the location of high harmonic sources.

Reporting: The 95th percentile value of each harmonic order, including THD, is the accepted reporting indices. In some cases, e.g. IEC 61000-3-6, 99th percentile and maximum values are reported.

8.3.4 Voltage fluctuations and flicker

Definition: Voltage fluctuations and flicker whilst linked to the same source are not interchangeable. Voltage fluctuations are a cause whilst flicker is an effect. Voltage fluctuation is defined as "a series of voltage changes or a continuous variation of the RMS voltage" [6]. Flicker is defined as "impression of unsteadiness of visual sensation induced by a light stimulus whose luminance or spectral distribution fluctuates with time" [12].

Limits: Compatibility limits for flicker are specified in IEC 61000-2-2 [13] and are 1 for P_{st} and 0.8 for P_{lt}. IEC 61000-3-7 [14] also specifies indicative planning levels for flicker in higher voltage networks as presented in Table 8.8.

There are no planning levels stated for LV systems, with limits being applied to emissions from equipment. However, MV planning levels are often carried through to the LV system.

Measurement: IEC 61000-4-30 specifies a methodology for measurement of rapid voltage change (voltage fluctuations). In the case of flicker, measurement is to be determined from the cycle-by-cycle RMS values using

Table 8.8 IEC 61000-3-7 flicker indicative planning levels [15].

	Planning levels	
	MV	HV-EHV
P_{st}	0.9	0.8
P_{lt}	0.7	0.6

the complex flickermeter algorithm described in IEC61000-4-15 [11]. For each 10-min interval, the algorithm gives a value for the short-term flicker severity index P_{st}, where $P_{st}=1$ corresponds to threshold of irritability for 50% of the population. For each 2-h period, the P_{st} values are combined into a long-term flicker severity index P_{lt}, defined by

$$P_{lt} = \sqrt[3]{\frac{1}{12}\sum P_{st}^{3}} \tag{8.4}$$

The calculation of flicker requires very fine voltage resolution and significant instrument processing power.

Reporting: The 95th percentile values of both P_{st} and P_{lt} over the measurement period are the generally accepted indices. Flicker compliance is generally assessed by comparing the 95th percentile values of P_{st} and P_{lt} collected over a 1-week period to the applicable limit.

8.3.5 Voltage sags

Definition: "A voltage sag or dip is a short term (less than 1 minute) reduction in RMS voltage below 90% of its nominal value" [16].

Limits: There are no internationally accepted limits for voltage sags.

Measurement: From the RMS value of each cycle, the voltage envelope can be determined. Power quality monitors generally have adjustable sag or dip threshold settings, corresponding to the level of the horizontal line in Fig. 8.2. The sag start and finish times are determined as the instants at which the RMS voltage falls below and then increases above the threshold, respectively, allowing the sag duration to be determined. There is often a small margin known as hysteresis allowed to ensure that the voltage has clearly returned to within the normal operating range before the event trigger is rearmed. The sag voltage is the magnitude of the retained voltage (often given as a percentage of the nominal voltage) during the sag.

The earlier description has been applied to a single-phase sag. It becomes more difficult to characterise a sag in a 3-phase system when there are different voltage envelopes on the phases. The solution to this problem is time and phase aggregation. Phase aggregation involves aggregating sags occurring at a common time stamp down to one sag voltage and duration. Time aggregation involves aggregating all sags occurring within 1 min of each other into one sag; this prevents sags caused by recloser operations due to the initial fault being reported as discrete sags. Time and phase aggregation is described in detail in Chapter 6.

Reporting: Other than reporting each sag individually (huge amounts of data), there is no universally accepted method of reporting sags. However, there are many standards and academic papers which suggest indices for reporting of sags. In terms of standards, IEEE 1564 is the IEEE Guide for Voltage Sag Indices. This standard specifies indices for single events, sites and systems. Further detail on reporting of sags is provided in Chapter 6.

8.3.6 Transients

Definition: A transient is a subcycle disturbance in the AC waveform that is evidenced by a sharp, brief discontinuity of the waveform. Transients are also known as spikes or surges. There are two types of transients, namely, oscillatory (capacitor switching) or impulse (lightning). In the case of impulsive transients they are normally present for only 1/1000th of a second or less (less than 1 ms). Oscillatory transients are generally present for up to ½ a cycle. Transient magnitudes can be from a few to 10,000 V-peak above or below the voltage sine wave.

Limits: There are currently no widely accepted limits for transient.

Measurement: Of all power quality disturbances, characterisation, analysis and reporting of transients are least developed. Some instruments may give a log of transients above a set threshold voltage with no other information. Others may give the energy contained in the transient, assuming a standard load impedance.

Reporting: At present, there are no well accepted methods of reporting transients.

8.4 Power quality instruments

8.4.1 Instrumentation requirements

An instrument for measuring PQ disturbances is called a power quality monitor. Power quality monitors must be able to identify and record the characteristics of many types of disturbances changing on a timescale of microseconds (transients) to hours (steady-state voltage variations). Fast transients require high sample rate analogue-to-digital converters (e.g. 1–4 MHz) giving a large data throughput. Operating over a long timescale this gives an enormous amount of data to be handled. These instruments must either record very little of the data handled or have large storage facilities (e.g. hard disk or compact flash) or communicate the data to a storage facility by means of a modem or integrated communications systems such as local area networks and fibre.

Power quality monitors may have to meet stringent standards for measurement and processing methods. Since transients require the measurement of voltages larger than the peak supply whilst flicker requires the resolution of voltages of 0.3% or less, their input must have a wide dynamic range with accurate analogue-to-digital converters (e.g. 12-bit converters used in some monitors are not accurate enough for flicker).

Table 8.9 summarises the instrumentation requirements for the different PQ disturbance types. Voltage, harmonics, sags and swells are relatively standard features. Unbalance requires three voltage channels and direct measurement of line–line voltages or measurement of the phase of the line–neutral voltages. Measurement of voltage fluctuations requires expensive analogue-to-digital converters and complex processing. It is only available

Table 8.9 Summary of instrumentation requirements.

Disturbance	Logging/ event capture	Special requirements		Comments
		Voltage	Frequency	
Voltage	Log		50 Hz	
Unbalance	Log	3 phase	50 Hz	Line–line voltages required
Harmonics	Log	FFT to 20-40th	Maybe 2 kHz	
Fluctuation	Log	0.1% resolution	50 Hz	Complex processing required if irregular
Sag/interruption	Event	Threshold	50 Hz	Complex for 3-phase circuits and when several close events occur
Swell	Event	Threshold	50 Hz	
Transient	Event	Threshold	5 kHz–5 MHz	Desirable to separate impulsive and oscillatory

in expensive instruments. Transients also require expensive analogue-to-digital converters.

The following considerations are required when choosing a power quality monitor:

1. They are available in a range of price points (from a few hundred to several thousand dollars) and performances. It is necessary to understand performance requirements well.
2. Which monitoring standards does the instrument comply with if any?
3. It is desirable in general that they can measure 3-phase voltage and preferably also current, especially if they are to be used in fault finding.
4. The monitor should have sufficient memory capacity to store all required data for the survey period or be able to be interrogated before memory fills. Modern instruments have memory capacities of up to several gigabytes using compact flash cards or internal hard drives.
5. There should be adequate filtering, surge protection and battery backup such that the device is unaffected by disturbances (e.g. harmonic distortion, loss of supply) on their own supply leads.
6. It is most important that the monitor can give trends for voltage and harmonics, with unbalance and fluctuation as desirable but more expensive options. Ninety-five percent values should be available from the instrument or it should be possible to export the trend data to third party software for subsequent processing.
7. The instrument should be able to recognise and characterise sags, interruptions and swells. There should be adjustable thresholds for each.
8. The instrument should at least give a count of transients above a known threshold level.
9. For fault finding, it is essential that it can store some waveforms with time and date stamp.
10. The monitor should have a ruggedised case if it is to be installed in an uncontrolled environment and be weatherproof if used outdoors.
11. The reporting software should be easy to use and flexible. Flexibility is not so important if it is easy to export the data to other analysis packages, such as Microsoft Access, Microsoft Excel or Mathworks Matlab.

Fig. 8.4 illustrates an example specification for an Australian designed and manufactured modern power quality monitor.

8.4.2 Instrumentation used for power quality monitoring

For the purposes of this book, instruments capable of measuring power quality to some degree have been classified into the following four main types:

PARAMETER	DESCRIPTION
Power quality parameters	
Class A declared/nominal input	230V 50Hz/60Hz
Power frequency	IEC61000-4-30 (section 5.1).
Magnitude of the supply voltage	IEC61000-4-30 (section 5.2).
Flicker	IEC61000-4-30 (section 5.3).
Supply voltage dips and swells	IEC61000-4-30 (section 5.4).
Voltage interruptions	IEC61000-4-30 (section 5.5).
Supply voltage unbalance	IEC61000-4-30 (section 5.7).
Voltage harmonics	IEC61000-4-30 (section 5.8).
Voltage interharmonics	IEC61000-4-30 (section 5.9).
Mains signalling voltage on the supply voltage	IEC61000-4-30 (section 5.10).
Rapid voltage changes (RVC)	IEC61000-4-30 (section 5.11).
Underdeviation and Overdeviation	IEC61000-4-30 (section 5.12).
Measurement	
A to D Conversion	16 bits.
Samples per cycle	384 @ 50 Hz; 320 @ 60 Hz.
Sampling Rate	Nominal: 19.2kHz synchronised to mains.
Anti-aliasing	High-frequency components attenuated by at least 50dB so as not to interfere with harmonic measurements.
Measurement metrics	
Frequency	Range: 50Hz nominal (42.5-57.5)Hz, 60Hz nominal (51.0-69.0)Hz; Full range: (40-70)Hz. Measurement: 10s; Accuracy: ±5mHz referenced to RTC, ±1mHz referenced to GPS.

Fig. 8.4 Example specification for a power quality monitor.

- Handheld/spot check
- Smart tariff meters
- Mid-range power quality
- High-range power quality

Each of these four main types of instrument is defined as follows.

8.4.2.1 Handheld/spot check instruments

Some simple PQ investigations can be made without a fully equipped power quality monitor. Handheld instruments include the following:

- Multimeter: Used to perform simple checks on voltage and unbalance levels. Many can also measure current. It is important that true RMS multimeters be used for power quality purposes.
- Portable harmonic meter: Useful when harmonics are not time varying; some can store and download to a personal computer and are useful in time-varying situations.

Handheld instruments are often very accurate; however, they generally have limited or no logging capability. They are generally significantly cheaper than specialised power quality monitors. Handheld meters are also limited in the types of disturbances that can be measured, e.g. handheld meters generally cannot measure flicker. As a general principle, handheld instruments will not comply with power quality monitoring standards such as IEC 61000-4-30.

Fig. 8.5 shows a photograph of an example of a handheld instrument with some power quality measurement functionality. In the case of this device, measurement of harmonics is possible. However, it does not have the capability to measure disturbances such as flicker and voltage sags. The ability to record and store data is also limited and not compliant with standards such as IEC 61000-4-30.

8.4.2.2 Smart tariff meters

Smart tariff meters are essentially electricity metering devices with some added power quality functionality. They are often seen as a fiscally attractive option for routine power quality monitoring as they are generally installed at installations for metering purposes and are significantly less expensive than power quality monitors. Adding power quality functionality generally does not present a major incremental cost. Many network operators are now using smart tariff meters to comply with some regulatory requirements for power quality monitoring. Fig. 8.6 shows a photograph of a smart revenue meter with harmonics measurement functionality.

Fig. 8.5 Photograph of an example of a handheld instrument with power quality measurement functionality.

Fig. 8.6 Photograph of an example of a smart revenue meter with harmonics measurement functionality.

Testing has shown that smart tariff meters perform well for basic parameters such as fundamental voltage and current and low order harmonics. The drawback of smart tariff meters is that they generally don't comply with monitoring standards and instrument accuracy becomes a major problem at higher order harmonics. In addition, the functionality of smart tariff meters is often limited with regard to the amount of on-board memory, the number and types of disturbances that can be measured and the sampling rates.

8.4.2.3 Mid-range power quality monitors

Mid-range power quality monitors are instruments that are designed for power quality monitoring that may not comply with monitoring standards. These are generally older (pre IEC 61000-4-30) or cheaper devices. Whilst they often measure the full spectrum of power quality disturbances they are generally most effective for measurement of voltage and current and low order harmonics. These devices can also be weatherproof and can be used outdoors. However, they can be limited in functionality, memory and sampling rates.

Fig. 8.7 Photograph of an example of a mid-range power quality monitor.

Mid-range power quality monitors are effective for surveys where high accuracy and repeatability are not paramount issues. Fig. 8.7 shows a photograph of a mid-range power quality monitor. This device is able to measure the majority of power quality disturbances but is not fully compliant with IEC 61000-4-30. This device also features a rugged waterproof case allowing it to be used in outdoor locations such as being mounted on a pole.

8.4.2.4 High-range power quality monitors

High-range power quality monitors are those which are fully compliant with modern power quality monitoring standards. They will monitor all power quality disturbances and have large amounts of on-board memory. They are very accurate and have fast sampling rates and extensive configuration options. In some cases, additional A/D converters are provided to give higher sampling rates for transient capture. Fig. 8.8 shows a photograph of an example of a high-range power quality monitor. This device is compliant with IEC 61000-4-30 Class A and will measure all power quality

Fig. 8.8 Photograph of an example of a high-range power quality monitor.

disturbances to the requirements of the standard. This instrument includes two A/D converters. The 'standard' A/D converter has sampling frequency of 200 kHz, whilst the second 'high speed' A/D converter has a sampling rate of 2 MHz. The high speed A/D is triggered by detection of a transient which exceeds the user-defined event threshold.

One of the drawbacks of high-range power quality monitors is that they are expensive. They can also be very complex to setup and instal and have complex software. Most cannot be used outdoors as they are not weatherproof.

8.4.2.5 Other instruments

Digital oscilloscopes (preferably with FFT feature) are useful for measurement of continuous disturbances under steady state and possibly under time-varying conditions as well depending on memory requirements. Flexible triggering modes and very high sampling rates make them very useful for capture of events such as switching transients. Fig. 8.9 shows a photograph of a digital oscilloscope.

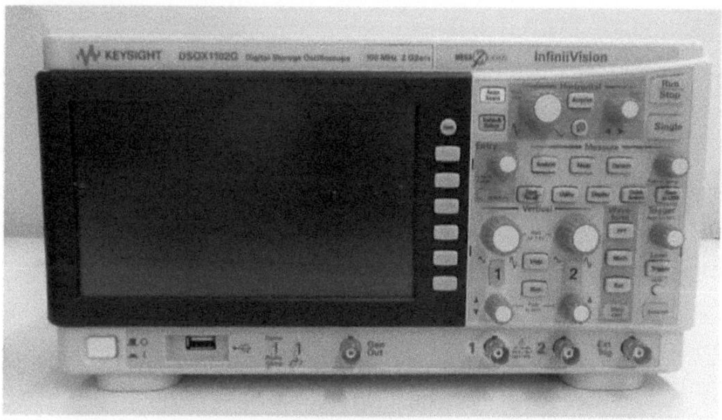

Fig. 8.9 Photograph of a digital oscilloscope.

Another class of modern instrument continually logs waveform data over time. These devices sample and store waveforms over long time periods using large on-board data storage (many GBs) in conjunction with data compression algorithms. Analysis and reduction of data for reporting is completed using software after the data has been captured. These devices have the advantage that events will not be missed due to incorrect trigger settings. In addition, it is sometimes useful to capture many cycles of waveform and these devices accomplish this task much better than traditional PQ monitors.

8.5 Transducers

Transducers play a very important role in power quality monitoring. Transducers are necessary to reduce the magnitude of the signal to be monitored down to levels which can be connected to the monitoring instrumentation. PQ monitors usually come equipped with voltage probes, and associated measurement hardware, which can be connected directly to LV circuits (generally up to 600-V RMS). For higher voltages, a transducer must be used to reduce the voltage to a value which can be accommodated by the monitor, for example, the secondary voltage of a substation metering transformer.

Investigation of particular customer problems usually requires measurement of current as well as voltage. Some instruments may directly measure small current inputs (e.g. 5 A). Newer instruments often use current transducers which convert current signals to voltage inputs.

The accuracy of the transducers used to perform power quality measurements can have a significant impact on the measurement outcomes. For harmonics, the frequency response of the transducer must be adequate up to the frequency of the harmonic of interest. Transient measurement also requires frequency response adequate to resolve the transient without the loss of important detail.

8.5.1 Classes of transformer

There are two main classes of transformer (voltage or current) found in the field (e.g. substations) that are relevant to power quality monitoring: measuring and protective. It is important to understand the distinction between these transducer classes and selection of transformer class can have significant implications on power quality measurement outcomes. IEC 60050 [17] defines measuring and protective transducers as follows:

- Measuring voltage transducers—voltage transformer intended to transmit an information signal to measuring instruments, integrating metres and similar apparatus.

- Protective voltage transformer—voltage transformer intended to transmit an information signal to electrical protective and control devices.

The definition for current transformers is as mentioned before with 'current' swapped for 'voltage'.

IEC 61869 [18] provides standard accuracy classes for voltage transformers. These are presented in Table 8.10.

Measuring transformers are designed to have high accuracy but have a limited measurement range. Protective transformers have lower accuracy but a much larger measurement range before saturation in order to accommodate the large values that can be associated with power system faults and transients. In practical terms, this means that wherever possible measuring transformers should always be used for measurement of variations (as can be seen in Table 8.10, protective VT accuracy is much lower than that required for instrumentation compliant with IEC 61000-4-30 Class A as presented in Table 8.5). If it is expected that measurement of values which exceed the nominal by a large margin, protective transformers may be considered to be useful as measuring transformers are likely to saturate which can lead to clipping or other distortion of measurements.

8.5.2 Voltage transducers/transformers

Voltage transducers are used to reduce the voltages which are to be measured down to levels which are compatible with the capabilities of the monitoring instrument. Typically the instrument input voltage capability is below 600-

Table 8.10 Accuracy requirements for measuring and protection voltage transformers.

		Phase displacement	
Class	**Voltage ratio error (±%)**	**± Minutes**	**± Centiradians**
Measuring voltage transformers			
0.1	0.1	5	0.15
0.2	0.2	10	0.3
0.5	0.5	20	0.6
1.0	1.0	40	1.2
3.0	3.0	Not specified	Not specified
Protective voltage transformers			
3P	3.0	120	3.5
6P	6	240	7.0

Reproduced from IEC, Instrument transformers—Part 3: Additional Requirements for Inductive Voltage Transformers, IEC 61869-3:2011, 2011.

V RMS. There are four main voltage transducer technologies in use in the field as follows:

- Inductive (magnetic) voltage transformers—these are common at voltages up to 132 kV.
- Capacitive voltage transformers (CVT)—these are common at voltages above 132 kV.
- Capacitive or resistive dividers—these are less common in the field but are often used for laboratory measurements.
- Electro-optic voltage transducers—these are newer technology and not prevalent in the field at present.

Each of the technologies mentioned before is described in detail as follows.

8.5.2.1 Inductive voltage transformers

Magnetic voltage transformers use the same principle as power transformers to reduce the voltage to the appropriate level for input to protection and metering devices. Fig. 8.10 shows the basic operating principle of an ideal magnetic voltage transformer. Fig. 8.11 shows photographs of a number of magnetic voltage transformers installed at substations.

The frequency response of an inductive VT depends on the type, manufacturer and the burden. The voltage level of the transducer also affects the measurement accuracy of the transducer. At high voltage the transformer frequency response may only be adequate up to several hundred Hertz. Fig. 8.12 shows the ratio correction factor (value by which the VT output must be multiplied in order to correct for variation) response for three sample VTs. In this figure, the nominal voltage rating of the three samples is similar; Sample 1 has a rating of 27.5 kV whilst samples 2 and 3 are rated at 26.4 kV. This figure clearly shows that VTs with similar voltage ratings can have frequency response which differs significantly. In this example,

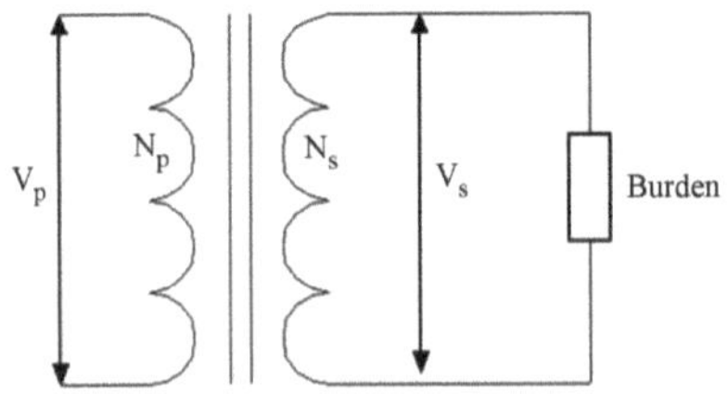

Ignoring Voltage Drop

$$V_s = \frac{N_s}{N_p} \times V_p$$

- V_p – Primary voltage
- N_p - Number of turns on primary winding
- V_s – Secondary voltage
- N_s - Number of turns on secondary winding

Fig. 8.10 Operating principle of an ideal magnetic voltage transformer.

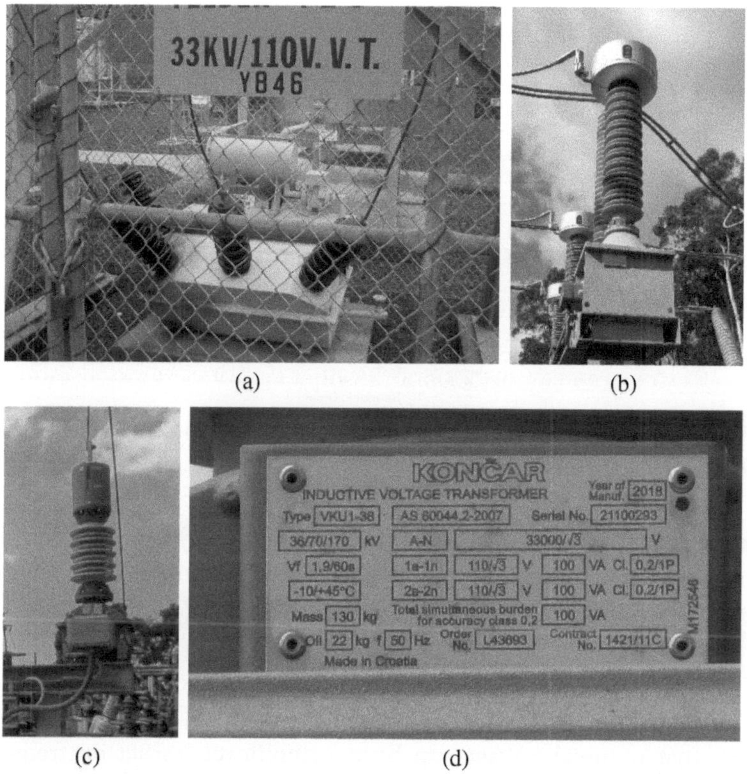

Fig. 8.11 Photographs of inductive VTs (A) three-phase 33 kV:110 V, (B) single-phase 132 kV:110 V, (C) single-phase 33 kV:110 V, (D) single-phase 33 kV:110 V nameplate.

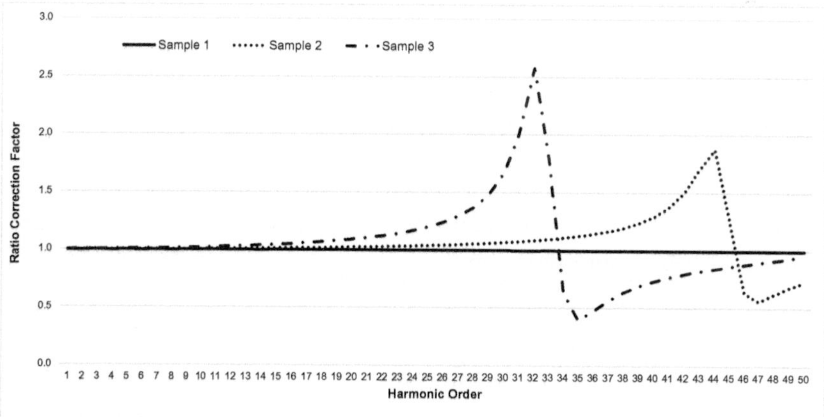

Fig. 8.12 Ratio correction factors for three sample VTs.

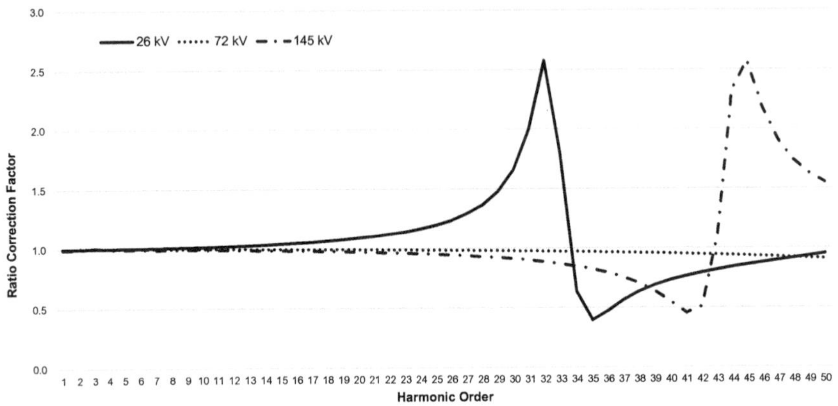

Fig. 8.13 Ratio correction factors for three VTs showing impact of nominal voltage rating.

Sample 1 has an almost flat frequency response out to the limit of harmonic orders evaluated whilst the limit of measurement for Sample 2 is likely to be limited to harmonics out to approximately the 33rd order. For Sample 3, the limit of measurement is likely the 20th order.

Fig. 8.13 shows an example of how frequency response, as expressed by ratio correction factor, can vary based on VT nominal voltage. Here it can be seen that nominal voltage can have a significant impact on frequency response with the limit of measurement generally decreasing as nominal voltage increases. Finally, Fig. 8.14 shows the impact of burden on frequency response. This graphic demonstrates the highly variable nature of the frequency response of VTs. It can be seen that for one sample, burden has a significant impact whilst for the other sample, burden has almost no impact at all.

The examples of VT frequency response shown in this section demonstrate that it can be very hard to predict the frequency response of any VT without type testing information. Of the examples shown before, accurate harmonic measurement could only be guaranteed to approximately the 20th order. There is only one example where accurate harmonic measurements would be possible to the 50th order which is the traditional limit of measurement for power quality monitors.

8.5.2.2 Capacitive voltage transformer

The basic capacitive voltage transformer (CVT) combines a voltage divider with an inductive VT [also known as an electromagnetic unit (EMU)]. CVTs are used extensively at higher voltage levels (> 132 kV). Use of the divider at high voltage greatly simplifies the design of the inductive VT,

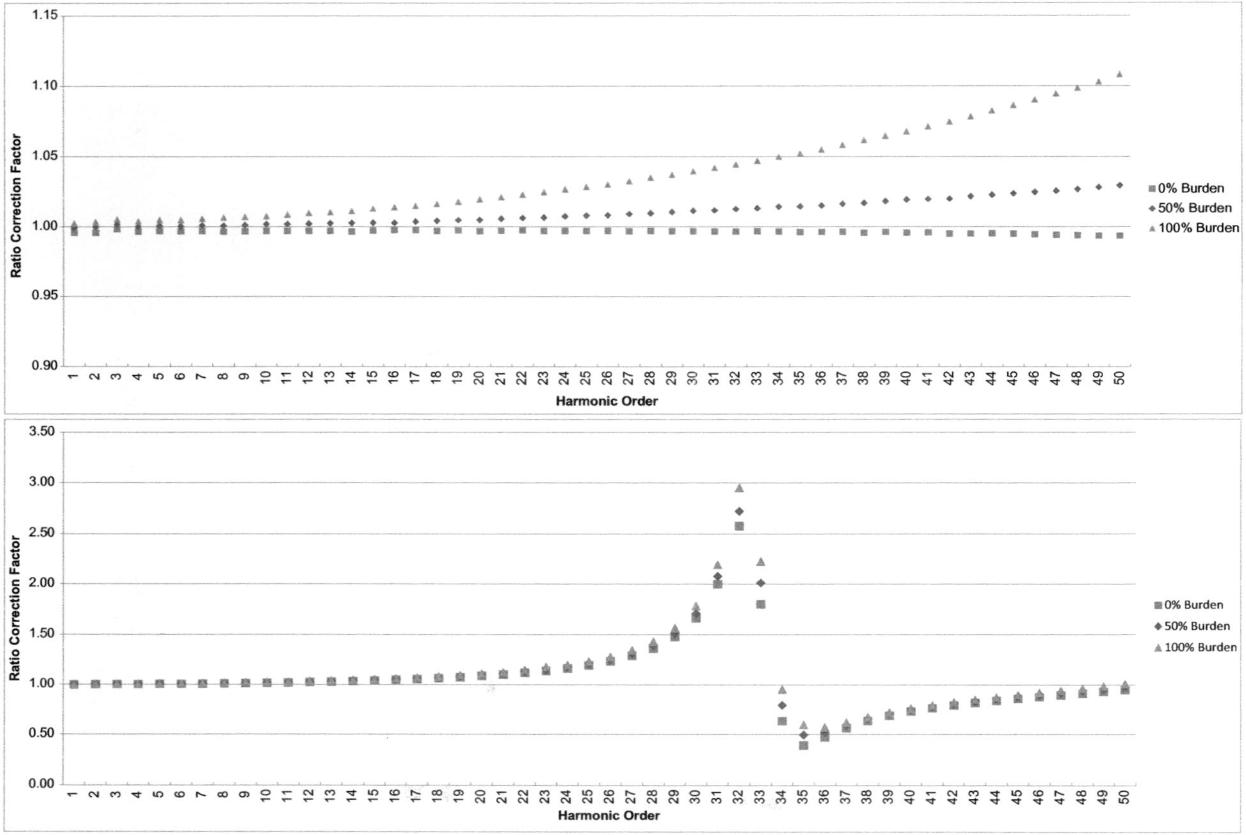

Fig. 8.14 Ratio correction factor for two VTs indicating impact of burden on frequency response.

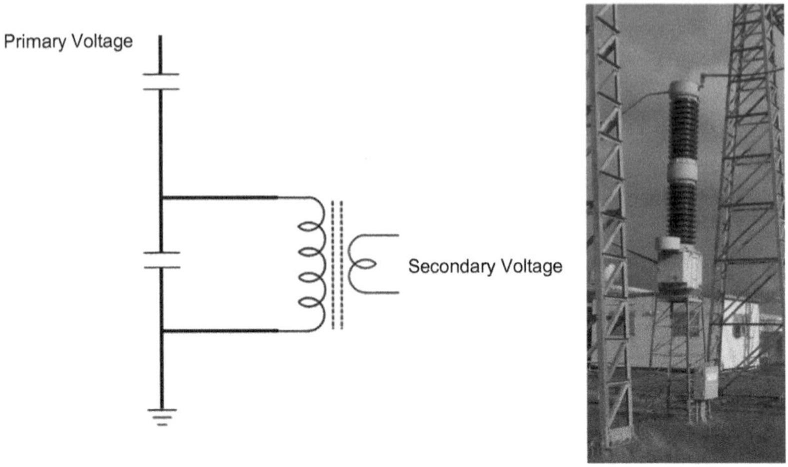

(a) (b)

Fig. 8.15 Operating principle of an ideal capacitive voltage transformer (A) and outdoor CVT (B).

making it smaller and more economic at HV. The capacitive divider and EMU are usually tuned close to the fundamental frequency. This greatly limits frequency response and as such unmodified CVTs are unsuitable for measurement of harmonics. Fig. 8.15A shows the operating principle of an ideal CVT. Fig. 8.15B shows a photograph of an outdoor-type CVT.

Whilst a basic CVT is not suitable for measurement of harmonics, there are a number of vendors who offer technology solutions which can allow use of CVTs for harmonics measurements.

8.5.2.3 Voltage dividers

Voltage dividers use the voltage divider principle to reduce voltages. Fig. 8.16A shows the operating principle of an ideal voltage divider.

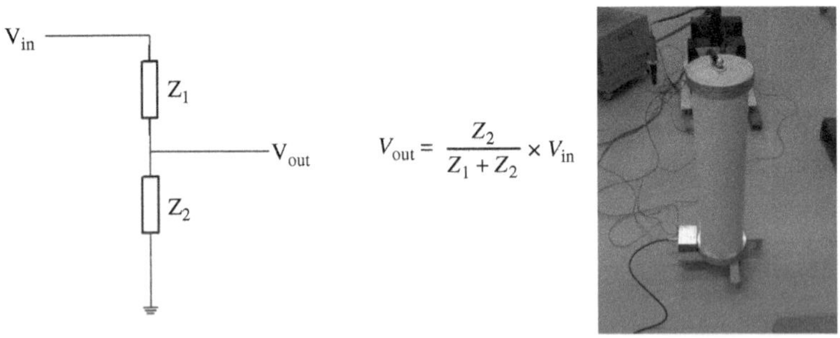

(a) (b)

Fig. 8.16 Operating principle of an ideal voltage divider (A), and laboratory capacitor voltage divider rated at 100-kV line-earth (B).

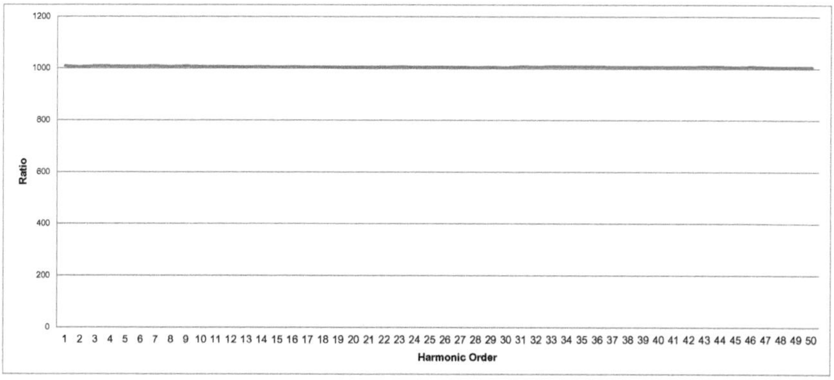

Fig. 8.17 Frequency response of 100-kV capacitive divider.

Z_1 and Z_2 may be resistors or capacitors or some combination of both and are specified to give the appropriate output voltage. Z_1 must be rated for the full voltage whilst Z_2 can be rated at a lower voltage. Devices to be connected to the voltage divider are connected between V_{out} and earth. Fig. 8.16B shows a photograph of laboratory capacitor voltage divider rated at 100-kV line-earth.

Voltage dividers generally have very high-frequency bandwidth making them very suitable for measurement of harmonics. However, relatively few voltage dividers are found in service in the field. Fig. 8.17 shows a frequency response measurement for the capacitive divider shown in Fig. 8.16. The nominal ratio for the divider is 1000. As can be seen, there is little variation in the divider ratio out to the 50th harmonic.

8.5.2.4 Electro-optic voltage transducers

Electro-optic voltage transducers are a relatively modern technology and are not widely used at present. Electro-optic voltage transducers use the Kerr effect, also called the quadratic electro-optic effect, which is a change in the refractive index of a material in response to an applied electric field. Electro-optic voltage transducers should have high-frequency response.

8.5.3 Current transformers

A current transformer (CT) can be defined as "an instrument transformer in which the secondary current, in normal conditions of use, is substantially proportional to the primary current and differs in phase from it by an angle which is approximately zero for an appropriate direction of the connections." Current transformers reduce large current to values suitable for input to protection and measurement devices. In many cases, particularly more recently, CTs may be incorporated in switchgear. There are four main current transformer technologies in use:

- Inductive (iron-cored)
- Rogowski (air-cored)
- Hall effect
- Magneto-optic

Each of the above is described in detail as follows.

8.5.3.1 Inductive current transformers

Fig. 8.18 shows the operating principle of an inductive CT. The current induced in the secondary winding of the CT is proportional to the current in the primary conductor.

Fig. 8.19 shows an example of an inductive CT installed in a 132-kV switchyard.

For low voltage applications, inductive CTs may be window, wound or bar type whilst at high voltage designs include hair pin, eye bolt and top core. Many of the clamp-on type CTs, used by power quality monitors, example shown in Fig. 8.20, use the same principles as an inductive CT. However, these clamps on CTs have additional circuitry which converts the induced current to a voltage signal. In the example shown, the clamp on CT will have an output of 1 mV per ampere. The principle of operation for inductive type CTs, including the clamp-on CT shown, is such that they are only capable of measuring AC currents (i.e. DC current cannot be measured using these devices).

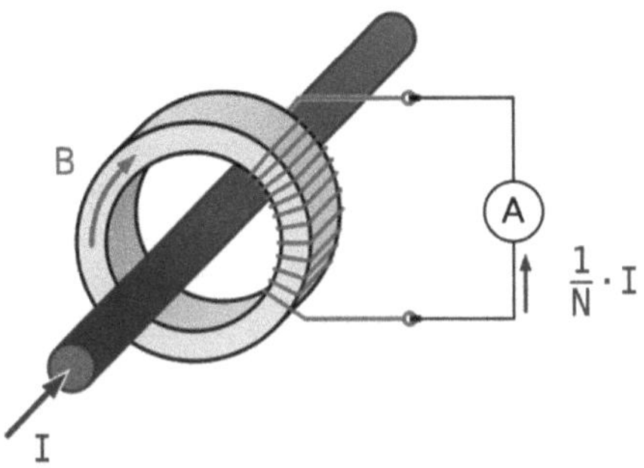

Fig. 8.18 Inductive CT principle of operation. *(© Biezl, This file is licensed under the Creative Commons Attribution 3.0 Unported licence.)*

(a) (b)

Fig. 8.19 Photograph of CT, (A) CT, (B) nameplate.

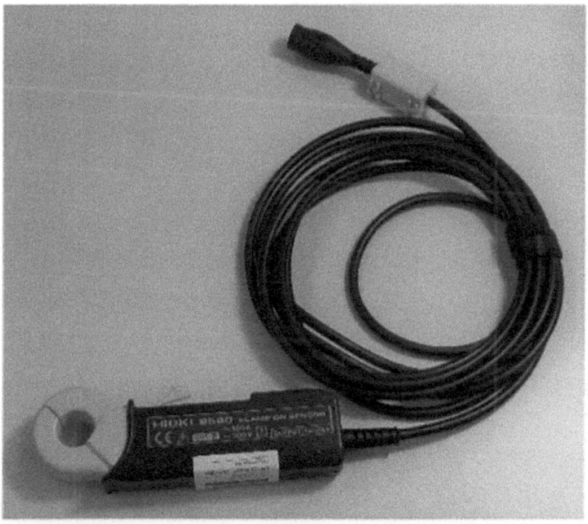

Fig. 8.20 Photograph of clamp-on CT used by Hioki power quality monitors.

Tests have shown that normal CTs are suitable for most harmonic measurements up to 2 kHz. However, specialised CTs are available which have enhanced frequency response.

8.5.3.2 Rogowski current transformers

A Rogowski coil is a toroid of wire used to measure an alternating current through a cable encircled by the toroid. The current in the cable encircled by the toroid produces a voltage in the toroid of wire. The output voltage of the coil is connected to an integrator circuit to obtain an output voltage that is

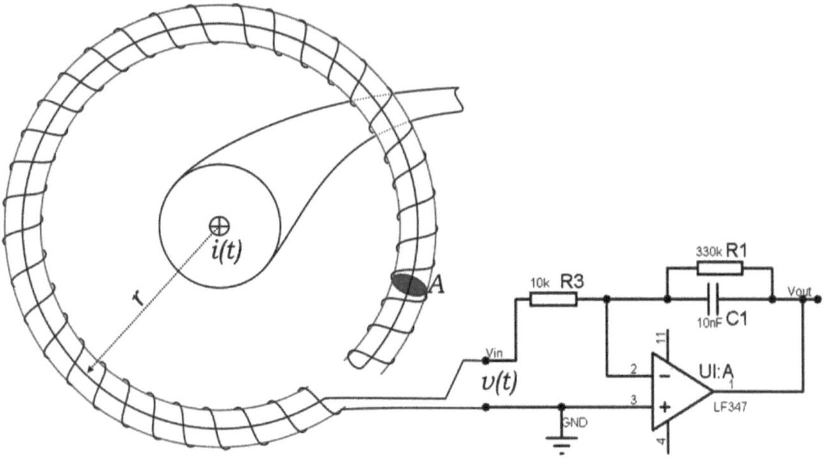

Fig. 8.21 Operating principle of a Rogowski coil. *(© Luque alfredo, This file is licensed under the Creative Commons Attribution 3.0 Unported licence.)*

proportional to the current in the cable. Fig. 8.21 shows the operating principle of a Rogowski coil.

Rogowski coils, example shown in Fig. 8.22, are used extensively in power quality monitoring. Rogowski coils have a number of advantages in this application, including that they can measure across a wide current range and they have a wide aperture allowing them to be place around cables with a much larger circumference than can typically be tolerated by clamp-on CTs. However, a disadvantage is that Rogowski coils require an external power supply for the integrator circuit.

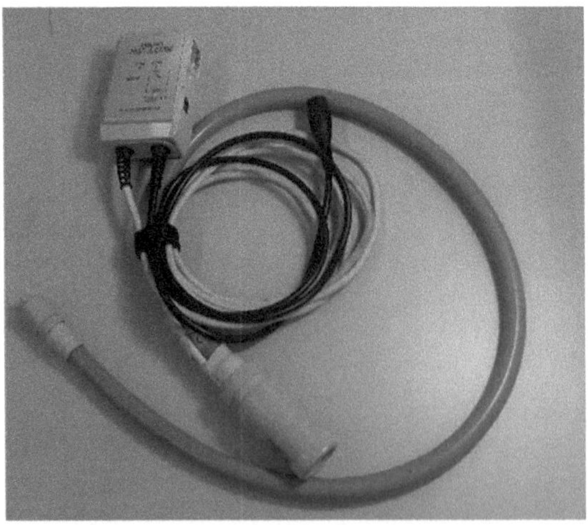

Fig. 8.22 Photograph of a Rogowski coil used by Hioki power quality monitors.

8.5.3.3 Hall effect current transformers

Hall effect current transformers use a Hall effect sensor which is a device that measures magnetic field. The output of a Hall effect sensor is directly proportional to the magnetic field strength through it. Due to their operating principle, Hall effect sensors are able to measure DC currents—something inductive and Rogowski current transformers are unable to do. Fig. 8.23 shows a photograph of a Hall effect CT, including the power supply required for the CT to operate.

8.5.3.4 Magneto-optic current transducer

Magneto-optic current transducers apply the magneto-optic Faraday effect. The magneto-optic Faraday effect describes the phenomenon that polarised light rotates under the influence of the magnetic fields and the rotation angle is proportional to the strength of the magnetic field component in the direction of optical path. Whilst the theory behind these transducers is

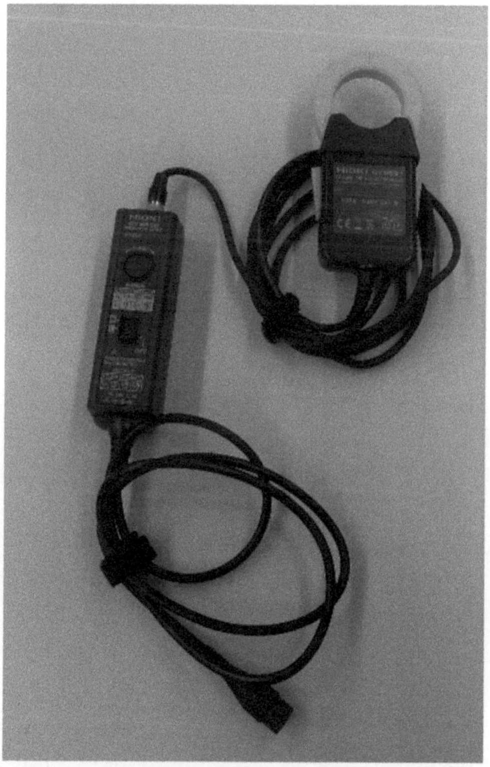

Fig. 8.23 Photograph of a hall effect clamp on CT used by Hioki power quality monitors.

not new, practical applications are only relatively recent and they are not commonly found on the field.

8.6 Motivation for power quality monitoring

As discussed in the introduction, there are two distinct motivations for power quality monitoring: reactive monitoring and proactive monitoring. Reactive monitoring is conducted in response to a known abnormal operation of the electricity system often arising from either a customer complaint (in the case of the network operator) or maloperation or failure of equipment. Proactive monitoring refers to routine monitoring usually performed by network operators. Proactive monitoring can be used for a variety of reasons, including regulatory requirements and for planning purposes. This section discusses reactive and proactive monitoring in detail.

8.6.1 Reactive power quality monitoring (fault finding)

Reactive monitoring may be undertaken by the network operator or the customer and is usually the result of a specific equipment problem or failure. Reactive monitoring often involves monitoring of both the voltage and current and often requires waveform capture to diagnose problems.

8.6.1.1 Fault finding methodology

Basic methodology of PQ fault finding involves determining answers to the key questions as follows:

i. **What equipment is affected?** Is the equipment being affected by a power quality disturbance problem (i.e. via the supply) or some other problem such as overheating or vibration? Is the problem permanent (damage has probably been caused) or intermittent (malfunction whenever a certain power quality disturbance magnitude is reached)?

ii. **What is the disturbance type?** What part of the equipment is likely to be the 'weak link'? What are the critical sensitivities of the equipment, e.g. harmonics, sags? If the problem is intermittent, a key step is to correlate maloperation with power quality levels on the supply. A monitor should be installed for a long enough period for the fault to appear several times. The times of fault occurrence should be checked against recorded power quality magnitudes. It is not sufficient to check against discrete disturbances. Continuous trends should also be checked since they might have high peaks at certain times. It may also be necessary to check other quantities since some power quality problems might only occur when there is a combination of high levels with high temperatures.

iii. **Is it within allowable limits?** When several correlations have been made between equipment maloperation and power quality disturbance magnitude, sufficient information should be available to allow determination of what power quality disturbance and magnitude thereof will cause problems and, of equal importance, what magnitude will be satisfactory. It is then necessary to determine whether the problem is an emission problem (too high a magnitude on the supply) or an immunity one (equipment tolerance is too low). This is a relatively easy question where standards are well defined, e.g. supply voltage magnitude, unbalance, harmonics and fluctuation, but requires some engineering judgement otherwise (e.g. voltage sags, transients). If the problem lies with the equipment, it needs to be redesigned for greater immunity or local mitigation installed. The remaining questions focus on the other possibility of the problem, that being power quality disturbance magnitudes which are too high on the supply.

iv. **From where does the disturbance originate?** The key step is to associate a problem appearing as a voltage disturbance with a current disturbance due to either a load or some other interaction. It may help to be able to disconnect the installation and monitor the incoming supply to see if the disturbance is still there (this allows identification of whether the disturbance is being generated by the local load or is external). It should be reported to the network operator when the supply is suspected as the point of origin.

v. **Should the fix occur at source, path or victim?** The terms source, path and victim are standard electromagnetic compatibility ones and based on a well-known model for how one electrical device affects another via the supply connections. To some extent, the answer to this question depends on whether the problem originates in the installation experiencing the problem or not. If it originates external to the installation, the network operator should determine the most cost-effective means of mitigating the problem. If both source and victim occur in the installation, cost comparisons should reveal where modifications need to be made. To take one extreme, if a cheap item of equipment is affecting many expensive items, the fix should be made at the source or possibly in the path (e.g. by filtering). There are other situations where the source is expensive, and the only problem is one piece of equipment needing a capacitor to be replaced by something larger—for example where sags affect a small power supply.

Table 8.11 Likely causes of common power quality problems.

Symptom	Likely cause
Capacitors fail	Harmonics
Clock resetting	Sag
Destruction of electronic equipment	Impulsive transients
Light fluctuation	Voltage fluctuations
Overloaded neutral	Harmonics
Racing clock	Harmonics

8.6.1.2 What type of disturbance to look for?

Table 8.11 summarises some common PQ problems and gives the most likely cause. It should be noted that UPS systems have been found to be a common cause of PQ problems if they are not properly designed and installed. One problem is that they can wrongly sense that the supply is faulty and spuriously switch in and out. Their waveform may not be sinusoidal for several cycles after switching and cause errors and resetting in digital equipment.

8.6.2 Proactive power quality monitoring

Whilst reactive monitoring may be undertaken by either a network operator or a specific installation, proactive monitoring is generally the domain of the network operator. In the past, wide-scale PQ monitoring was done infrequently, usually in response to special circumstances. Major motivations were as follows:

- In the distribution system: to assist customers experiencing PQ problems.
- In transmission systems: to establish existing values when planning for the installation of large distorting loads (e.g. smelters, renewable energy farms).

Power quality monitoring activity has increased considerably over the past decade. Some network operators now provide customers with typical values of power quality disturbances on their systems, for customer goodwill and to protect themselves from litigation, especially from customers who have been supplied with equipment with poor power quality immunity. Regulators have also begun to require that network operators define their power quality objectives. It is important that any power quality values used to describe network performance come from the network itself rather than be copied from figures for other networks with different design, construction and maintenance practices. Proactive monitoring can aid in this. Regulators have also required network operators to institute regular monitoring and reporting procedures so that the level of power quality disturbances can be shown objectively. In addition, in recent times, the proliferation of large renewable energy generators has increased the need for power quality visibility in order to undertake informed planning functions (e.g. emission allocation calculations) and to assess network compliance.

Knowledge of power quality levels within networks has other immediate benefits. It gives feedback as to voltage levels and fault impact, thus aiding asset management decisions. It is suggested that examining the trends in power quality disturbance levels will help identify network deficiencies before they become a problem. As network operators incorporate power quality considerations into their planning strategies, power quality database interrogations to identify trends can be used to check if the power quality planning methodology is effective and guide its further development.

Before proactive power quality monitoring can be undertaken it is necessary to develop some idea of what the expected outcomes of the project should be and how these will be achieved. Proactive monitoring should aim to give a broad understanding of network power quality levels. With this outcome in mind, the following are important theoretical questions which needed to be addressed before any proactive power quality monitoring project can get underway:

- What disturbances should be measured and why?
- Should monitoring be performed at low voltage or medium voltage?
- How should sites be selected?
- What should the survey duration be?
- How is the data to be retrieved from remote instruments?
- Where and how is the vast amount of data generated by proactive monitoring to be stored?
- How can this data be reduced, for reporting purposes, to an amount which is understandable without losing any important detail?

Each of the points mentioned before is addressed as follows.

8.6.2.1 What disturbances are measured and why

The following disturbances have been identified as those which give the most beneficial information with the least amount of data. Other disturbances such as flicker, voltage swells and transients may also be of interest if the infrastructure (instrumentation and data handling facilities) is available or if there are known sources of these disturbances (e.g. arc furnaces and flicker).

- **Voltage variation**: This is a basic power quality parameter and important for the operation of equipment. Overvoltage may cause equipment damage whilst undervoltage may cause equipment to maloperate or overheat.
- **Voltage unbalance**: High levels of voltage unbalance can cause additional heating in three-phase motors.
- **Voltage harmonics**: Harmonic distortion is an important power quality parameter as it is known to cause the maloperation of some equipment (such as certain types of relays), causes additional losses in motors due to counter-rotating magnetic fields and may destroy improperly installed power factor correction capacitors.
- **Voltage sags**: These very short duration changes in voltage magnitude are due to system faults and may be the most costly of all power quality disturbances to customers. Depending on the severity of the voltage sag, equipment may trip out resulting in significant economic loss, especially for some critical continuous production processes such as plastic extrusion.

In many proactive power quality monitoring schemes only voltage disturbances are measured as these provide indication of overall power quality levels as seen by the customer. Current measurement is also generally more difficult than voltage measurement and most useful for investigations of specific problems as opposed to routine (proactive) monitoring.

8.6.2.2 Should monitoring be at low voltage or medium voltage?

Monitoring of power quality at high or medium voltage will give a good indication of the power quality levels seen by large customers and one site may cover supply to many customers (e.g. a zone substation). However, the bulk of customers are connected to the low voltage network and this is where power quality disturbances will be at their worst. Thus, for an overall view of network power quality, it is necessary to monitor at both high, medium voltage and low voltage levels.

8.6.2.3 Site selection

Selection of sites should be made so that a good indication of power quality levels across the entire network is achieved. Consideration should be given to ensuring that the site selection methodology includes sites across the range of characteristics likely to be encountered in the network. Examples of site characteristics include the following:

- **Voltage level**—most networks will consist of a number of voltage levels. Measurement at medium and high voltage will allow a high-level understanding of the power quality levels seen by a large number of customers but in many cases will not identify the power quality disturbance levels seen by a customer in a low voltage network.

- **Site strength**—for the purposes of power quality monitoring, site strength is related to the impedance of the network and in turn related to the location of the site in relation to the nearest transformer. The fundamental classification is 'strong' and 'weak'. A strong site is one which is close to the supply transformer such as zone substation busbars or distribution transformer terminals at low voltage sites. Weak sites are those towards the end of feeders in the medium voltage case or at the end of distributors in the low voltage case. As an example, general methodology for determination of whether a site is strong or weak could be that if the fault level at the site is more than 50% of the fault level at the busbar of the nearest transformer, the site is considered to be strong. If the fault level is less than 50% of that at the busbar of the nearest transformer, the site is classified as weak. Fig. 8.24 shows a graphical representation of a method to characterise strong and weak sites.
- **Load**—four classifications for load type are suggested: industrial, commercial, residential and mixed. The mixed network type would be applied to sites such as zone substations which supply a varied range of customer types.
- **Network type**—the network topology can have significant impacts on power quality disturbance magnitudes. For example, factors such as network exposure can have major implications of voltage sag levels. Examples of network classifications for a distribution network are as follows:

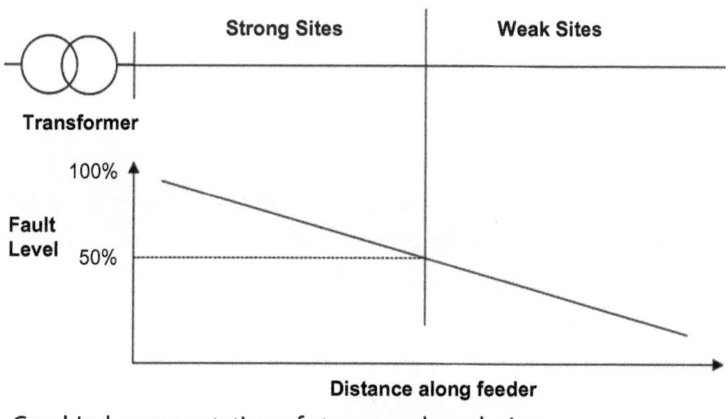

Fig. 8.24 Graphical representation of strong and weak sites.

- City—these are strong networks which have short feeders, are predominately underground and can be highly meshed.
- Urban—these networks have a combination of overhead and underground feeders up to 10 km in length.
- Short Rural—these networks are predominately overhead with feeders with length of tens of kilometres.
- Remote—these networks consist of very long overhead feeders.

Fig. 8.25 shows possible locations for power quality monitors within a power system. Whilst this graphic is able to capture site characteristics such as voltage level, load type and site strength, it is not able to capture the network topology characteristics which may be equally important to the characteristics shown.

Another important aspect of developing a proactive power quality monitoring system is selection of the number of sites which require monitoring. Whilst at high voltage and medium voltage, the number of substations and customer connections may be small enough to allow monitoring at all or most connection points, this will not be the case at low voltage where there are many thousands or perhaps millions of connection points in each network. IEC documents such as [19] favour an approach which involves 95% compliance in time and space. Put more simply, this means that 95% of sites should comply 95% of the time. However, 95% of available LV sites is an impractically large number of sites to directly monitor. This means a sample of the population of LV sites should be selected and results analysed using statistical estimation techniques. Whilst there is very limited literature available which gives guidance as to the sample size required to prove network compliance at LV sites, the Council of European Energy Regulators (CEER) Guidelines of Good Practice on the Implementation and Use of Voltage Quality Monitoring Systems for Regulatory Purposes [20] recommends the following site numbers for various statistical indicators of overall network performance:

- 20 sites if averages over all locations will be reported.
- 200 sites if 95th percentile values over all locations will be reported.
- 1000 sites if 99th percentile values over all locations will be reported.

Whilst the CEER guidelines do give specific site numbers, no reference is made as to how these were obtained. The study presented in Ref. [21] presents statistical analysis of existing data to provide some insight into the number of sites required for accurate estimation of the 95th percentile value the low voltage values of various power quality disturbances across an entire large network. The number of sites determined for voltage variation, voltage

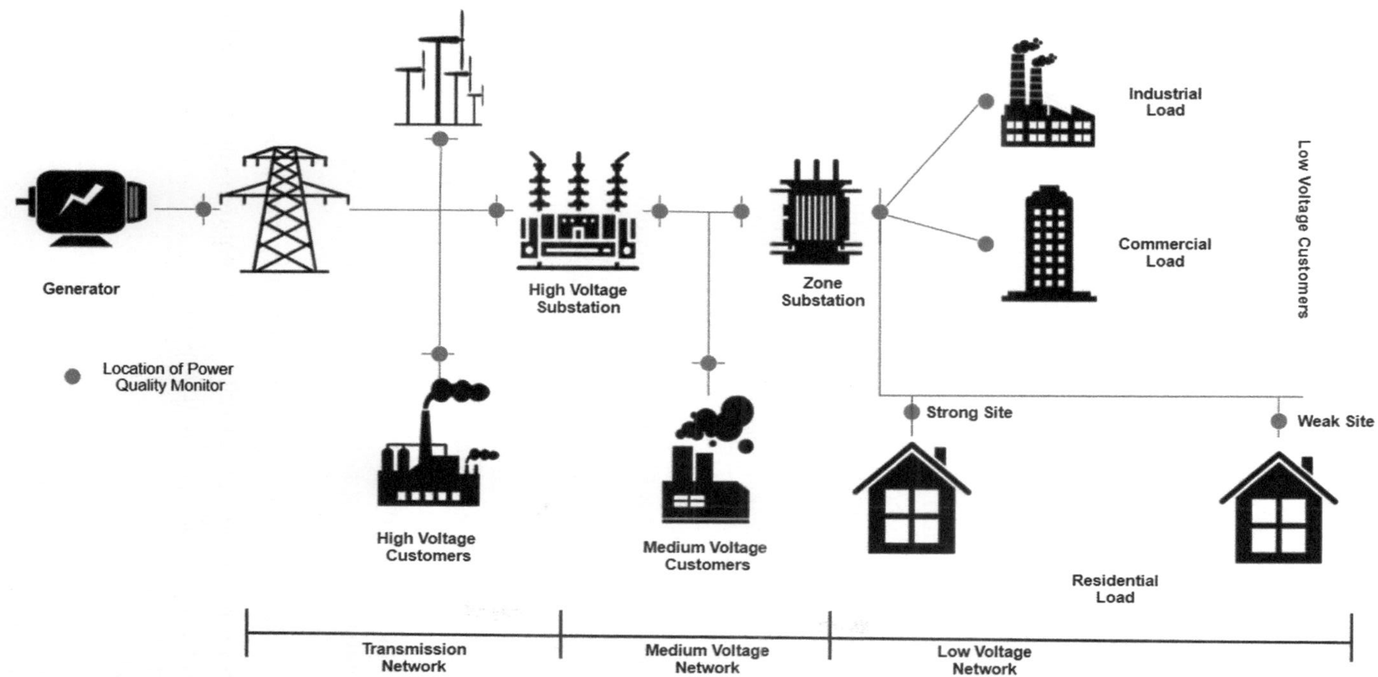

Fig. 8.25 Examples of possible power quality monitoring locations in a power system.

Table 8.12 Calculated number of sites to accurately estimate low voltage 95th percentile values across an entire network.

Disturbance	Number of sites
Voltage variation	100
Voltage unbalance	~500
Voltage harmonics	~200

unbalance and voltage harmonics (as represented by voltage THD) is presented in Table 8.12. Whilst this data gives some indication of the requisite site numbers, it should be noted that these values may only be valid for a single network.

8.6.2.4 Survey duration

For continuous disturbances (voltage variation, voltage unbalance and voltage harmonics), most standards call for a minimum survey duration of 1 week in order to determine compliance since this covers the normal business cycle. However, there are a number of studies, including [22], which tend to suggest that there can be significant seasonal variation in disturbance levels. This may not be a significant issue if permanent monitoring systems are installed; however, if nonpermanent monitoring is to be used, consideration should be given to allowing either sufficient duration to account for seasonal variation or short-term monitoring should be repeated at different times of the year.

For voltage sags, the survey period needs to be much longer as these disturbances are dependent on many random variables such as weather, bushfire activity and human interaction with the power system. Indeed, some sites may not experience any sags for a whole year but then several in 1 week. The accepted minimum period for effective voltage sag surveying is 1 year although in practice it may be several years or more before a definite trend becomes apparent.

8.6.3 Practical aspects of power quality monitoring

In practice, it is not possible to fulfil all the requirements outlined before. There can be limitations on the following:

- Instrumentation
- Site selection
- Transducers
- Data retrieval and storage

Each of the above is examined in detail as follows.

8.6.3.1 Instrumentation

Ideally, all power quality monitoring would be performed by specialised, power quality monitoring equipment compliant with the most stringent requirements of the relevant power quality standards. However, this equipment is expensive to buy and in the past it has been difficult for network operators to develop business cases to justify purchase of these devices. In practice, power quality data will likely be available from a variety of the instrument types defined in Section 8.4.

8.6.3.2 Site selection

In practice, it is very rare to have monitoring of weak sites due to the fact that most fixed installation power quality or smart revenue metres are installed at zone substations of large customers who have their own substations. Whilst it is noted that the penetration of smart revenue meters in low voltage networks is increasing rapidly, increasing access to data from weak sites, in general, the metres used only have capability to provide voltage magnitude measurements and do not have the functionality to measure other PQ disturbances. In addition, the measurement protocols used in these devices are not compliant with PQ monitoring standards.

8.6.3.3 Transducers

In practice, it may not be feasible to measure all of the disturbances that are of interest due to transducer limitations, as discussed in Section 8.5. Further, in the case of medium voltage it is difficult to find suitable transducers for measurement away from substations (such as at the end of feeders). Hence, at the present time, the type of sites monitored will probably be dominated by strong sites.

8.6.3.4 Data retrieval and storage

A power quality monitor at one site produces 3 measurements each 10 min for voltage. When harmonic measurements are taken up to the 40th or 50th harmonic, plus unbalance and flicker this can result in more than 100 parameters. If minimum, average and maximum values are recorded, this value increases by a multiple of 3. As such, it is clear that power quality monitoring can result in significant dataflows. A power quality survey conducted at hundreds of sites over a 1-year period yields huge amounts of data. This large volume of data needs to be collected from remote instrumentation and stored.

In order for the data to be retrieved from remote instrumentation, communications infrastructure is required that will facilitate this data transfer. In many cases, data transfer is achieved through the use of GSM modems. In some cases, power quality data is also collected using revenue metering data collection schemes. These techniques are not necessarily the optimum means of data transfer. Communications infrastructure has improved considerably over the last few years and some utilities now have fibre or other network (internet) type connections to many monitoring points.

Once the data has been retrieved it is then necessary to have the infrastructure to store it. Many power quality instruments now have proprietary databases associated with them to handle the data storage. Depending on individual requirements this may or may not be the optimum solution, particularly where a utility is using many different types of instruments to obtain power quality measurements. Some proprietary databases also do not give the flexibility that may be required for effective power quality assessment.

8.7 Reporting of power quality data

As stated before, routine (proactive) power quality monitoring can produce very large volumes of data. A fundamental challenge of reporting power quality data is to reduce this large volume of data down to a form which is understandable without loss of important detail. As a general principle, beyond specifying methods of compliance assessment for individual sites, standards do not provide prescriptive methodologies for analysis and reporting of large volumes of power quality data. This section provides guidance with respect to methods that can be used to analyse and report the large volumes of data which will be generated by proactive power quality monitoring surveys. The focus here is on reporting of variations. Methods of reporting voltage sag data are dealt with in detail in Chapter 6.

When considering methods of reporting large volumes of data, it can be useful to consider a tiered reporting approach. Taking the example of a network operator, there will be a range of management levels for which reporting needs to be tailored. For example, a senior manager is unlikely to be interested in the detailed performance of an individual site. Instead, they will more likely want to understand the overall performance of the networks and perhaps how it compares with their peers (benchmarking). Conversely, the power quality engineer may be interested in the detailed performance of sites as it is their job to rectify issues. Use of a tiered reporting approach can generate reports that are useful across the range of business requirements. Again using the example of a network operator, three tiers of reporting can be identified as follows:

- Site reporting—this is the most detailed tier of reporting and includes visualisation of raw data for each site.
- Network reporting—this tier involves reporting of individual site performance through the use of a limited amount of indices for each site.
- Utility reporting—in this tier a few summary indices or graphics which describe the high-level performance of the entire network are presented.

Fig. 8.26 shows a graphical representation of a tiered power quality reporting structure.

8.7.1 Site reporting

The site reporting tier contains the most detailed data. Raw data which has not been processed into indices can be visualised using two main methods: time series trends and histograms.

8.7.1.1 Time series trends

Time series trends are useful as they can allow identification of periods where disturbance levels may be high or otherwise enable identification of trends over time. However, the usefulness of time series trends diminishes as the time period over which data is shown increases due to difficulties displaying large volumes of data. Fig. 8.27 shows an example of a time series trend of flicker values over a 1-week period. The time series trend allows identification of the movement in disturbance levels over time but generally will not

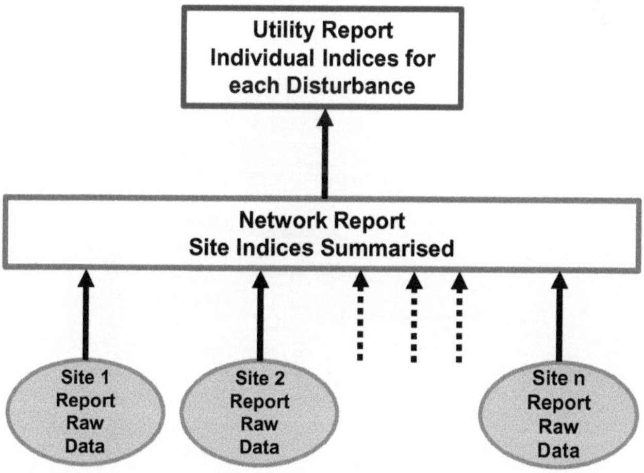

Fig. 8.26 Tiered reporting structure for power quality data.

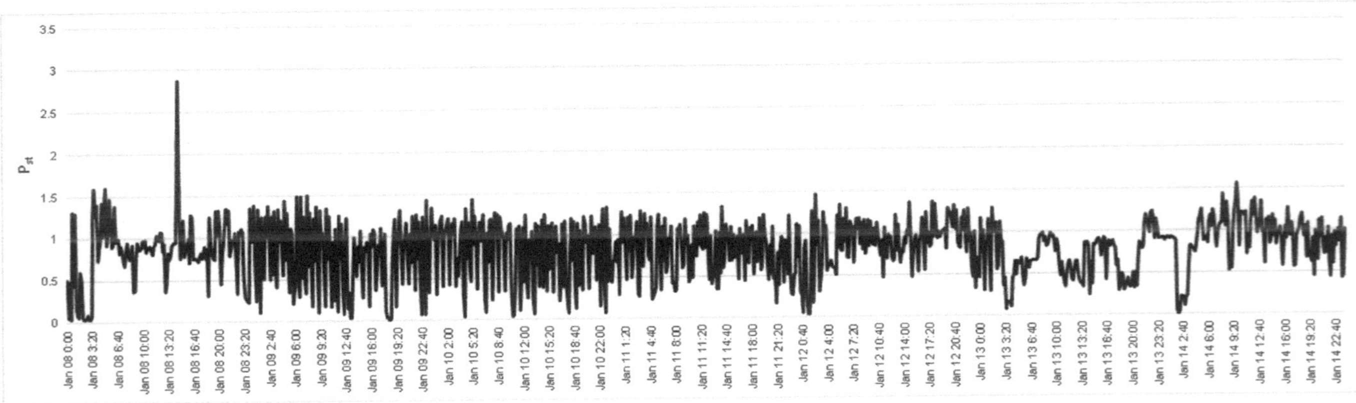

Fig. 8.27 Example of a time series trend.

allow statistical analysis and as such will not allow assessment of compliance. In addition, it will become more and more difficult to discern behaviour as the timeframe increases.

8.7.1.2 Histograms

Histograms are a useful method of showing large volumes of data as the histogram is unaffected by the amount of data to be displayed. Histograms can be overlaid by limit lines allowing quick identification of the amount of data which may exceed a limit. However, it is not possible to identify any time-related trends in performance from a histogram. Fig. 8.28 shows an example of a histogram of voltage unbalance measurements at a site. In this particular histogram, a line indicating the limit is shown. The histogram allows the user to quickly identify the distribution of disturbance magnitudes at the site and also to assess what proportion of the data is above the limit.

8.7.2 Network reporting

The network reporting tier uses indices calculated for each site and can provide some indication of the ranking of sites from worst to best (based on a ranking metric). For the purposes of the network reporting tier, it is necessary to calculate site indices from the raw data. The most common method of

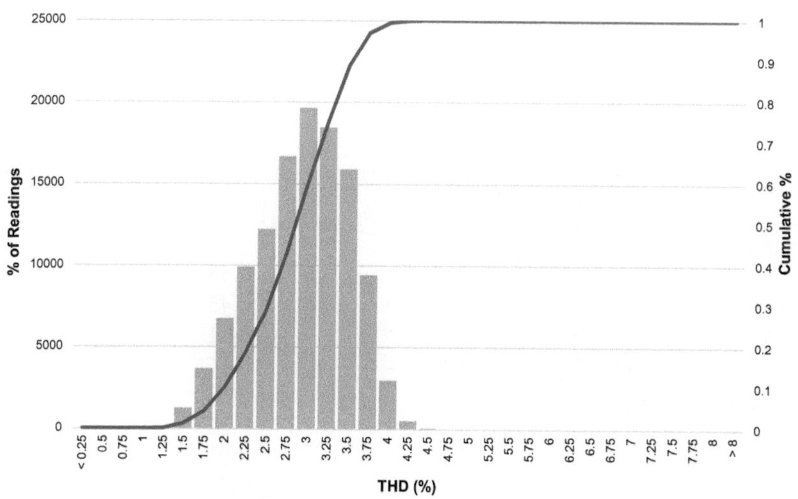

Fig. 8.28 Example histogram for harmonics measurements.

calculating a site index for variations is to perform statistical analysis on the data. In many cases, analysis and derivation of an index for power quality data involves the use of a statistical measure called a percentile value with the 95th percentile value being in common usage. As described in Section 8.1.4, a percentile value is the value which will not be exceeded for a given percentage of the data (which may or may not equate to time). For power quality analysis, the 95th percentile is often preferred over other statistical measures such as the average or the maximum, as the average value may not allow an adequate understanding of values at the upper end of the range (which may be the most important from an equipment impact perspective) whilst the maximum value may represent a transient occurrence that is not representative of ongoing network behaviour. The statistical metric selected for network reporting is somewhat arbitrary; however, it is often useful to select a metric which aligns with the metric used for compliance assessment. This then allows for easy identification of compliant and noncompliant sites within the network reporting tier.

Table 8.13 presents an example of a network report which consists of a table of site indices for flicker. As can be seen, sites have been ranked based on the value of the indices. These rankings can then be used to order sites as presented in Table 8.14. Ordering sites in this way allows quick identification of the poor performing sites.

Fig. 8.29 shows an example of a histogram of site indices. Whilst similar to the histogram shown in Fig. 8.28, this histogram allows easy identification of the percentage of sites which may exceed the limit. The histogram has

Table 8.13 Example of network reporting—table of site indices.

	95th percentile P_{st}		95th percentile P_{lt}	
Site	**Index**	**Site rank**	**Index**	**Site rank**
PL00001	0.22	6	0.21	7
PL00003	0.22	6	0.34	4
PL00018	0.26	2	0.45	2
PL00019	0.26	2	0.45	3
PL00022	0.19	9	0.21	8
PL00032	0.16	10	0.16	10
PL00045	0.10	11	0.08	11
PL00054	0.24	5	0.25	5
PL00055	0.22	8	0.20	9
PL00060	0.50	1	0.53	1
PL00066	0.25	4	0.23	6

Table 8.14 Example of network reporting—site indices ranked by disturbance magnitude.

Site	95th percentile P_{st}	Site rank
PL00060	0.50	1
PL00018	0.26	2
PL00019	0.26	2
PL00066	0.25	4
PL00054	0.24	5
PL00001	0.22	6
PL00003	0.22	6
PL00055	0.22	8
PL00022	0.19	9
PL00032	0.16	10
PL00045	0.10	11

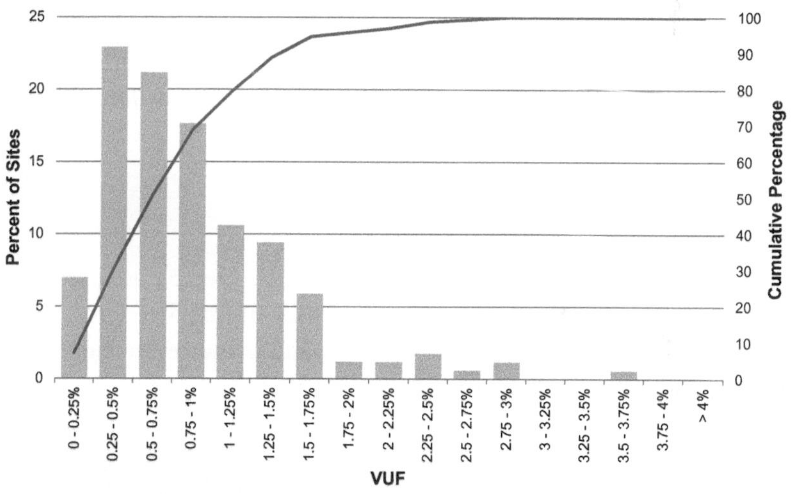

Fig. 8.29 Example site indices histogram.

advantages over tabular methods of reporting as the graphic can accommodate any number of sites and remain the same size.

8.7.3 Utility reporting

The utility reporting tier contains summarised site data designed to give an indication of performance across all of the sites provided by a participant, i.e., a high-level overview of all site indices for each PQ parameter. Examples of methods that can be used for utility reporting include:

- Compliance summaries
- Distribution of site indices
- Utility indices

8.7.3.1 Compliance summary

A compliance summary is a simple method of demonstrating the percentage of sites which are noncompliant. An example of a compliance summary graphic is shown in Fig. 8.30.

8.7.3.2 Distribution of site indices

A distribution of site indices graphic can be used to illustrate the performance of all sites for each power quality parameter in decile bands. Further insight can be gained by normalising site indices against the relevant limits. Fig. 8.31 shows an example of a distribution of sites graphic for a network operator. In this graphic, the upper extent of the bar for each disturbance shows the magnitude of disturbance index at the 95th percentile site whilst the lower extent of the bar shows the magnitude of disturbance index at the 5th percentile site. From this graphic it is easy to visualise the performance of the 95th percentile site with respect to the limit. The spread of disturbance level can also be easily visualised.

8.7.3.3 Utility indices

Utility indices are calculated based on all sites submitted by the participant. These indices can provide single numbers for each disturbance across all sites. Examples of statistical measure that can be useful as indices include median values (the 50th percentile) and 95th percentile values. The median value

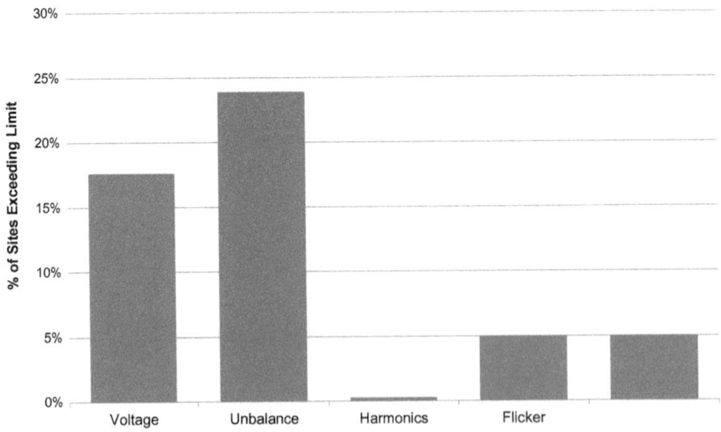

Fig. 8.30 Example summary of compliance graphic.

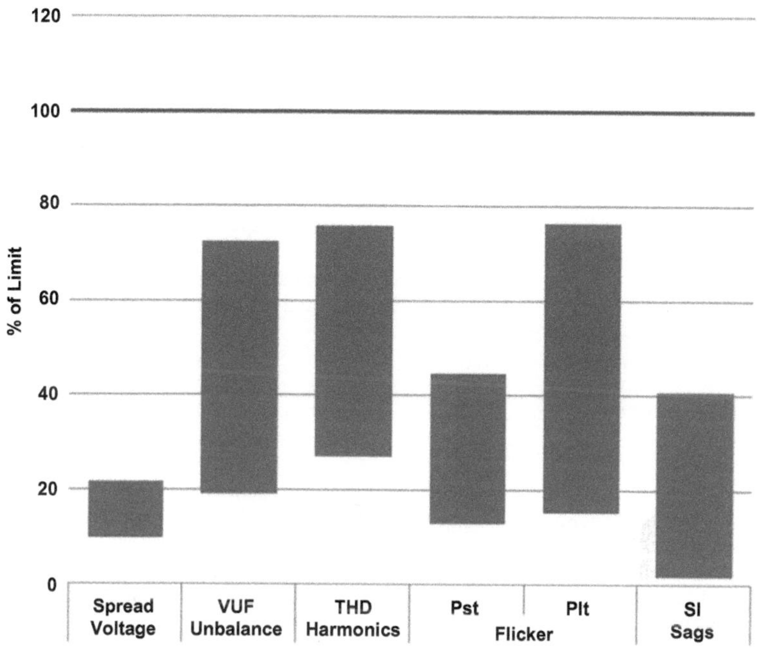

Fig. 8.31 Example distribution of site indices graphic.

can be used as a measure of the average performance of the participant whilst the 95th percentile value provides an indication of the disturbance levels that will be experienced at the worst 5% of sites. The value is calculated as the 95th percentile level of indices for sites. Table 8.15 presents a snippet of site indices and illustrates how a utility index can be calculated.

Table 8.15 Example calculation of utility index.

Site	95th percentile P_{st} index
PL00001	0.22
PL00003	0.22
PL00018	0.26
PL00019	0.26
PL00022	0.19
PL00032	0.16
PL00045	0.10
PL00054	0.24
PL00055	0.22
PL00060	0.50
PL00066	0.25
Utility median value	0.22
Utility 95th percentile value	0.38

References

[1] R.C. Dugan, M.F. McGranaghan, H. Wayne Beaty, Electrical Power Systems Quality, McGraw-Hill, New Jersey, 1996.

[2] IEC, Electromagnetic Compatability (EMC)—Part 4-30: Testing and Measurement Techniques—Power Quality Measurement Methods, IEC 61000-4-30 Edition 3 2015-02, 2015.

[3] IEC, Electromagnetic Compatibility (EMC)—Testing and Measurement Techniques—Flickermeter—Functional and Design Specifications, 2010.

[4] IEC, Electromagnetic Compatibility (EMC)—Part 4-7: Testing and Measurement Techniques—General Guide on Harmonics and Interharmonics Measurements and Instrumentation, for Power Supply Systems and Equipment Connected Thereto, IEC 61000-4-7 2009, 2009.

[5] CENELEC, Voltage Characteristics of Electricity Supplied by Public Electricity Networks, EN 50160:2010, 2010.

[6] IEEE, IEEE Recommended Practice for Monitoring Power Quality, 2009.

[7] IEC, Assessment of Power Quality—Characteristics of Electricity Supplied by Public Networks, IEC TS 62749, 2020.

[8] Standards Australia, Technical Report Electromagnetic Compatibility (EMC) Part 3.13: Limits—Assessment of Emission Limits for the Connection of Unbalanced Installations to MV, HV and EHV Power Systems, TR IEC 61000.3.13:2012, 2012.

[9] V.J. Gosbell, H.M.S.C. Herath, B.S.P. Perera, D.A. Robinson, Sources of errors in unbalance measurements, in: Australasian Universities Power Engineering Conference, AUPEC'02, Melbourne, 29 September - 2 October, 2002.

[10] Integral Energy Power Quality Centre, Technical Note No. 3, Harmonic Distortion in the Electric Supply System, 2000.

[11] IEEE, IEEE Recommended Practice and Requirements for Harmonic Control in Electric Power Systems, IEEE Std 519-2014, 2014.

[12] Australia/New Zealand Standard for Electromagnetic Compatibility (EMC)—Limits—Limitation of voltage changes, voltage fluctuations and flicker in public low-voltage supply systems, for equipment with rated current <=16 A per phase and not subject to conditional connection, AS/NZS 61000.3.3, 2006.

[13] IEC, Electromagnetic Compatibility (EMC)—Part 2-2: Environment—Compatibility Levels for Low-Frequency Conducted Disturbances and Signalling in Public Low-Voltage Power Supply Systems, IEC 61000-2-2:2002, 2002.

[14] IEC, Electromagnetic Compatibility (EMC)— Part 3-7: Limits—Assessment of Emission Limits for the Connection of Fluctuating Installations to MV, HV and EHV Power Systems IEC TR 61000-3-7, 2008.

[15] Standards Australia, Technical Report—Electromagnetic Compatibility (EMC)—Part 3.7: Limits—Assessment of Emission Limits for the Connection of Fluctuating Installations to MV, HV and EHV Power Systems, TR IEC 61000.3.7:2012, 2012.

[16] Integral Energy Power Quality Centre, Technical Note No. 4, Voltage Sag Measurement and Characterisation, 2001.

[17] International Electrotechnical Vocabulary, Chapter 604: Generation, Transmission and Distribution of Electricity—Operation, 1998.

[18] IEC, Instrument transformers—Part 3: Additional Requirements for Inductive Voltage Transformers, IEC 61869-3:2011, 2011.

[19] IEC, Electromagnetic Compatibility (EMC)—Part 3-6: Limits—Assessment of Emission Limits for the Connection of Distorting Installations to MV, HV and EHV Power Systems, IEC/TR 61000-3-6 Edition 2 2008-02, 2008.

[20] Council of European Energy Regulators (CEER) and Energy Community Rgulatory Board (ECRB), Guidelines of Good Practice on the Implementation and Use of Voltage Quality Monitoring Systems for Regulatory Purposes, 2012.

[21] S. Elphick, V. Gosbell, V. Smith, G. Drury, R. Barr, Assessing network compliance for power quality performance, in: 16th International Conference on Harmonic and Quality of Power, ICHQP 2014, Bucharest, Romania, 25–28 May, 2014.
[22] J. Meyer, P. Schegner, G. Eberl, Increasing the reliability of indices for power quality assessment in distribution networks, in: 13th International Conference on Harmonics and Quality of Power (ICHQP 2008), Wollongong, Australia, September 28–October 1, 2008, pp. 1–6.

Index

Note: Page numbers followed by *f* indicate figures, and *t* indicate tables.

F

G

H

I

Kolibri 360 GmbH
Ferdinand-Jühlke-Straße 7
99095 Erfurt, Deutschland
produktsicherheit@kolibri360.de